MCDP 6

Command and Control

U.S. Marine Corps

PCN 142 000001 00

DEPARTMENT OF THE NAVY
Headquarters United States Marine Corps
Washington, D.C. 20380-1775

4 October 1996

FOREWORD

This doctrinal publication describes a theory and philosophy of command and control for the U.S. Marine Corps. Put very simply, the intent is to describe how we can reach effective military decisions and implement effective military actions faster than an adversary in any conflict setting on any scale. In so doing, this publication provides a framework for all Marines for the development and exercise of effective command and control in peace, in crisis, or in war. This publication represents a firm commitment by the Marine Corps to a bold, even fundamental shift in the way we will view and deal with the dynamic challenges of command and control in the information age.

The Marine Corps' view of command and control is based on our common understanding of the nature of war and on our warfighting philosophy, as described in Fleet Marine Force Manual 1, *Warfighting* (to be superseded by Marine Corps Doctrinal Publication 1, *Warfighting*). It takes into account both the timeless features of war as we understand them and the implications of the ongoing information explosion that is a consequence of modern technology. Since war is fundamentally a clash between independent, hostile wills, our doctrine for command and control accounts for animate enemies actively interfering with our plans and actions to further their own aims. Since we recognize the turbulent nature of war, our doctrine provides for fast, flexible, and decisive action in a complex environment characterized by friction, uncertainty, fluidity, and rapid change. Since we recognize that equipment is but a means to an end and not the end itself, our doctrine is independent of any particular technology. Taking a broad view that accounts first for the human factors central in war, this doctrine provides a proper framework for designing, appraising, and deploying hardware as well as other components of command and control support.

This doctrinal publication applies across the full range of military actions from humanitarian assistance on one extreme to general war on the other. It applies equally to small-unit leaders and senior commanders. Moreover, since any activity not directly a part of warfighting is part of the preparation for war, this doctrinal publication is meant to apply also to the conduct of peacetime activities in garrison as well as in the field.

This publication provides the authority for the subsequent development of command and control doctrine, education, training, equipment, facilities, procedures, and organization. This doctrinal publication provides no specific techniques or procedures for command and control; rather, it provides broad guidance which requires judgment in application. Other publications in the command and control series will provide specific tactics, techniques, and procedures for performing various tasks. MCDP 5, *Planning*, discusses the planning side of command and control more specifically.

"Operation VERBAL IMAGE," the short story with which this publication begins, offers a word picture of command and control in action (done well and done poorly) and illustrates various key points that appear in the text. It can be read separately or in conjunction with the rest of the text. Chapter 1 works from the assumption that, in order to develop an effective philosophy of command and control, we must first come to a realistic appreciation for the nature of the process and its related problems and opportunities. Based on this understanding, chapter 2 discusses theories of command and control, looking at the subject from various aspects, such as leadership, information management, and decisionmaking. Building on the conclusions of the preceding chapters, chapter 3 describes the basic features of the Marine Corps' approach to command and control.

A main point of this doctrinal publication is that command and control is not the exclusive province of senior commanders and staffs: effective command and control is the responsibility of all Marines. And so this publication is meant to guide Marines at all levels of command.

C. C. KRULAK
General, U.S. Marine Corps
Commandant of the Marine Corps

DISTRIBUTION: 142 000001 00

Command and Control

Operation VERBAL IMAGE

Chapter 1.   The Nature of Command and Control

Chapter 2.   Command and Control Theory

Chapter 3. Creating Effective Command and Control

The Challenges to the System — Mission Command and Control — Low-Level Initiative — Commander's Intent — Mutual Trust — Implicit Understanding and Communication — Decisionmaking — Information Management — Leadership — Planning — Focusing Command and Control — The Command and Control Support Structure — Training, Education, and Doctrine — Procedures — Manpower — Organization — Equipment and Technology — Conclusion

Notes

Operation VERBAL IM-AGE

Scene: A troubled corner of the globe, sometime in the near future. The Marine expeditionary force prepares for an upcoming offensive.

2248 Monday: Maj John Gustafson had taken over as the regimental intelligence officer just in time for Operation VERBAL IMAGE. *Who thinks up the names for these operations anyway?* he wondered. This would be his first command briefing and he wanted to make a good impression. The colonel had a reputation for being a tough, no-nonsense boss—and the best regimental commander in the division. Gustafson would be thorough and by-the-numbers. He would have all the pertinent reports on hand, pages of printouts containing any piece of data the regimental commander could possibly want. He went over his briefing in his mind as he walked with his stack of reports through the driving rain to the command tent.

The colonel arrived, just back from visiting his forward battalions and soaking wet, and said, "All right, let's get started. S-2, you're up."

Gustafson cleared his throat and began. He had barely gotten through the expected precipitation when the colonel held up his hand as a signal to stop. Gustafson noticed the other staff officers smiling knowingly.

"Listen, S-2," the colonel said, "I don't care about how many inches of rainfall to expect. I don't care about the percentage of lunar illumination. I don't want lots of facts and figures. Number one, I don't have time, and number two, they don't do me any good. What I need is to know what it all *means*. Can the Cobras fly in this stuff or not? Will my tanks get bogged down in this mud? Don't read me lists of enemy spottings; tell me what the enemy's up to. Get inside his head. You don't have to impress me with how much data you can collect; I know you're a smart guy, S-2. But I don't deal in data; I deal in pictures. Paint me a picture, got it?"

"Don't worry about it, major," the regimental executive officer said later, clapping a hand on Gustafson's shoulder. "We've all been through it."

0615 Tuesday: The operation was getting underway. In his battalion command post, LtCol Dan Hewson observed with satisfaction as his units moved out toward their appointed objectives. He watched the progress on the computer screen before him. Depicted on the 19-inch flat screen was a color map of the battalion zone of action. The map was covered with luminous-green unit symbols, each representing a rifle platoon or smaller unit. If a unit was stationary, the symbol remained illuminated; when the unit changed location by a hundred meters, the symbol flashed momentarily.

Hewson tapped on a unit symbol on the touch screen with his finger, and the unit designator and latest strength report came up on the screen. Alpha Company; they should be moving by now.

"Get on the hook and find out what Alpha's problem is," Hewson barked. "Tell them to get moving."

With rapid ease he "zoomed" down in scale from 1:100,000 to 1:25,000 and centered the screen on Bravo Company's zone. Hewson prided himself on his computer literacy; *no lance corporal computer operator necessary for this old battalion commander*, he mused. Hewson was always amazed at the quality of detail on the map at that scale; it was practically as if he were there. That was the old squad leader in him coming out. He tapped on the symbol of Bravo's second platoon as it inched north on the screen.

No, they should turn right at that draw, he said to himself. *That draw's a perfect avenue of approach. Where the hell are they going? Don't they teach terrain appreciation anymore at The Basic School?*

"Get Bravo on the line," he barked. "Tell them I want second platoon to turn right and head northeast up that draw. *Now.* And tell them first platoon needs to move up about 200 yards; they're out of alignment."

Satisfied that everything was under control in Bravo's zone, Hewson scrolled over to check on Charlie Company. Back when he was a young corporal, some 22 years ago, this technology didn't exist. It was amazing how much easier command and control was today compared to his old squad leader days, how much more control there was now. He wondered if the junior Marines realized just how lucky they were.

0622 Tuesday: Second Lieutenant Rick Connors was feeling anything but lucky. Just past the mouth of a draw, he angrily signaled for second platoon to halt. Company was on the radio, barking about something. He was wet, he was cold; his rain top had somehow sprung a leak, and a stream of icy water poured down his spine. And on top of everything else, now this.

"Come again?" he said to his radio operator.

"Sir, Hotel-3-Mike says we're supposed to turn right and head up this draw," LCpl Baker repeated.

Damn PLRS, Connors cursed to himself. He had never actually seen a PLRS, that venerable piece of equipment having been replaced by a newer, lighter generation of position-locating system which attached to any field radio and sent an updated position report every time the transmit button was cued. But like all the more experienced Marines, he insisted on calling the new equipment by the old name.

"Up that draw," Connors repeated, as if to convince himself he had heard correctly.

"Hotel-3-Mike says it's an excellent avenue of approach, sir," Baker reported dutifully.

Connors studied the impenetrable web of thorny, interlocking undergrowth in the draw and snorted scornfully. *Maybe on somebody's computer screen it is,* he thought. *But on the ground it's not. Somebody at battalion must have his map on 1:25,000 again. So much for the decentralized mission control they told us about at TBS. What do they even need lieutenants for if they're going to try to control us like puppets?* He despised the prospect of hacking his way through the thick brush of the draw, especially when first squad had spotted what looked like an excellent concealed avenue of approach not 200 yards ahead. Of course, if he followed instructions, higher headquarters would be squawking about his slow rate of advance—there were no thickets of pricker bushes on a computer map. He could just imagine the radio message: *"What's taking you so long, 3-Mike-2? It's only an inch on the map."* And if he chose the other route they'd be on him in no time about disobeying orders. He cursed the PLRS again. But then he decided it wasn't the PLRS that was the problem; it was the way it was being used.

1118 Tuesday: A section of SuperCobra IIIs churned through the driving rain on its way back to the abandoned high school campus that served as an expeditionary airfield, returning from an uneventful scouting mission.

"I'll tell you what, skipper," 1stLt Howard Coble said from the front seat of the lead helicopter, "this soup isn't getting any better."

In fact, it was getting considerably worse, Capt Jim Knutsen decided as he piloted the buffeting attack helicopter. A squall was moving back in. Goo at 500 feet, visibility down inside a mile and worsening.

"I'm glad I'm not those poor bastards," Coble said, indicating a mechanized column on the muddy trail below them to starboard.

"You got that right," Knutsen said, not paying much attention.

Until Coble cursed sharply.

"Those aren't ours," Coble said. "Take a look, skipper. BMPs, T-80s."

Coble was dead right. What they were looking at was an enemy mechanized column, Knutsen guessed, of at least battalion strength. Probably more. His first instinct was to make a run at the column, but his intuition told him otherwise. Something was not right. Knutsen banked the Cobra away sharply to avoid detection, and his wingman followed.

What's wrong with this picture? Knutsen said to himself. The mission briefing had said nothing about enemy mechanized forces anywhere near this vicinity. The enemy had apparently used the cover of the bad weather to move a sizable force undetected through a supposed "no-go" area into the division's zone. Knutsen was familiar enough with the ground scheme of maneuver to know instantly that this unexpected presence posed a serious threat to the upcoming operation. *We got ourselves a major problem. These guys are not supposed to be here.*

His wingman's voice crackled over the radio: "Pikeman, did you see what I just saw at two o'clock?"

"Roger, Sylvester."

"We need to let DASC know about this," Coble said on the intercom.

Knutsen considered the problem. Reporting the sighting to the direct air support center would, of course, be the standard course. But because of the weather, they'd had trouble talking to the DASC all day; they couldn't get high enough to get a straight shot. In these conditions, he figured they were nearly a half hour from the field. And when he finally got the message through, he could imagine the path the information would take from the DASC before it reached the units at the front—and that was provided they even believed such an unlikely report. *DASC hell, we need to tell the guys on the ground*, he thought. *They might like to know about an enemy mech column driving straight through the middle of the MEF's zone. Forget normal channels.* Unfortunately he had no call signs or frequencies for any of the local ground units.

"Howie, find me some friendlies on the ground," he said. He radioed his wingman with his plan.

"Got somebody, skipper," Coble said shortly. "AAV in the tree line at nine o'clock. Got it?"

"Roger, I'm setting down."

1132 Tuesday: *You got to be crazy to be flying in this weather,* Capt Ed Takashima said to himself when he heard the sound of approaching helicopter rotors. He was twice amazed to see the Cobra appear low over the trees and settle into the clearing not a hundred yards away while its partner circled overhead. He hopped down from his AAV and jogged out into the clearing to meet the Marine emerging from the cockpit and was three-times astonished to recognize him as an old Amphibious Warfare School classmate.

"Knut-case," he said, pumping his friend's hand enthusiastically. "I should have known nobody else would be crazy enough to fly in this stuff. What the hell are you doing here?"

Knutsen quickly explained the situation and, when he was finished and saw Takashima's expression, said: "Don't look at me like I'm crazy, Tak."

Anybody else Takashima would have thought *was* crazy—or else completely lost—but not Knutsen. He had known Knutsen too long for that. Knutsen was too squared away.

"Give me your map, I'll show you," Knutsen said. "We're right here, right? And the enemy is right *there*, heading in this direction," jabbing the map and tracing the enemy movement.

As Knutsen had begun to diagram the enemy move, Takashima was already considering the situation. *With all the sensors and satellites and reconnaissance assets that support a MEF*, Takashima wondered, *how does an enemy mechanized battalion drive through the middle of our sector without being detected?* He remembered reading something somewhere about uncertainty being a pervasive attribute of war. *Chalk one up to Clausewitz's "fog of war,"* Takashima decided. Of course, Takashima knew, since it was a "no-go" area—and that meant that somebody up the chain had looked at the terrain and decided it was impassable—it would remain relatively unobserved. But how it had happened didn't matter: it had happened. What to do about it? That was the problem. *Six or seven clicks, tops*, he thought, looking at the map. Not much time. This changed everything. The original battalion plan would have to be scrapped; it was as simple as that. Takashima recognized that his original mission was overcome by events. He made his decision. The situation called for quick thinking, and quicker action. The objectives might change, but the overall aim remained the same. The ultimate object, Takashima knew, was to locate the main enemy force and attack to destroy it. That could still be the object; it would just have to happen a lot farther south than had been planned. If the battalion could make a 90-degree left turn in time, they might just pull it off. Now if he could just get battalion to go along with it . . . . he needed to talk to the battalion commander.

Knutsen had finished tracing the enemy movement, and his finger rested on the map, pointing at a small town called Culverin Crossroads.

"That's it then," Takashima said. "Culverin Crossroads."

"I hear you, Tak," Knutsen said. "You're thinking of that West Africa map ex we did last year at AWS, aren't you? The one where we wheeled the whole regiment and took the red force in the flank."

"Yeah, that's the one," Takashima said.

"What the hell; let's do it. I got enough fuel for maybe one pass. You want me to work them over, or don't you want them to know that we're on to them?"

"Let's wait and surprise them. Can you bring back some friends?"

What a kick, Knutsen thought. *A couple of captains standing in the middle of a muddy field in a downpour working out the beginnings of a major operation.* It reminded him of playing pick-up football as a kid and drawing improvised plays in the dirt.

"We'll be here," he said with a grin. "You'll recognize me—I'll be the one in front."

"See you then, Knut-case," Takashima said.

They shook hands, and Knutsen climbed back into the cockpit.

"Olsen!" Takashima bellowed at his radioman. "Try to get me battalion. I need to talk to the colonel direct."

1310 Tuesday: "General, the latest weather pictures are coming in," the lance corporal reported, the note of anxiousness unmistakable in his voice.

MajGen Harry Vanderwood doubted if there was a single Marine anywhere in the wing who did not recognize the significance that attached to the latest forecasts.

"I'll be right there, Marine," he replied.

No sooner had Vanderwood arrived in the tactical air command center than the MEF commander bustled in unannounced as he had a disconcerting habit of doing. You never knew when he was going to show up, or where, Vanderwood mused. Wing commander or mechanic on the flight line, you were never safe.

"Have you gotten the latest on the situation, Harry?" the MEF commander asked.

"As of the last 15 minutes, general," Vanderwood replied. "Not that I'm any smarter than I was before. I'd still like to know what the hell is going on."

"That makes two of us. I'd like to talk to those Cobra pilots myself."

"It's being arranged, general. They managed to take off on another sortie before we could grab them. Under terrible conditions, I might add. When they get back, I'm either going to give them a medal or a butt-chewing; probably both."

The MEF commander grunted. "How's the weather looking?" he asked.

"We're just in the process of pulling down the latest pictures from the weather satellite," Vanderwood said.

A large-scale map of the area of operations appeared on the large screen, color-coded to illustrate the precipitation forecast.

"No good news there," Vanderwood said. "Let's take a look at the incoming weather."

A broader map, much like a weather map on a television newscast, appeared on the screen. Heavy white blotches swept sputteringly across the screen from left to right.

"Freeze it right there," Vanderwood said, and the image stopped moving. "Good. That could be the break we're looking for. I figure in about 90 minutes we'll be able to get something

11

going. If this pattern holds, I plan to blot out the sun—what little sun there might be—with aircraft by 1500. Now all we need is to know what we're going to be attacking."

"How about cueing up the MEF situation package, and we'll see if we can't make some sense of this," the MEF commander said. "And see if we can get General Bishop on teleconference."

"Somebody ask the Top to come over here," Vanderwood said, meaning the intelligence chief.

"General, the division commander's away from the CP, but we're setting up video with the chief of staff," a Marine reported.

"Very well," the MEF commander said. He fully expected Bishop to be away from the command post; in fact, the division command post was the last place he'd expect to find the division commander in the middle of a battle.

The computer operator, Cpl Beale Davis, tapped quickly on his keyboard, and the wall-sized screen blinked, the weather map replaced by a situation map of the MEF's area of operations. From the menu across the top of the screen, he opened a "conference" window, and the division chief of staff appeared in a live video feed.

"How are you, Tom?" the MEF commander said.

"Hanging in, general," Col Tom Hester replied. "Sir, General Bishop has gone forward. Do you want him paged? If he's at one of the regimental CPs, we can get him on video too."

"No, that's all right. We're just going to try to piece this picture together, and I want everybody to share the same image. Are you looking at the same thing we are?"

"Yessir, he is," Davis said, meaning that the screen in the division command post would depict the same information and images that were being called up on the wing situation map.

Davis had logged into the theater data base and could "pull down" almost instantaneously any individual piece of data, or complete or partial package of information, that had been entered into the system anytime, anywhere, by any means. He had access to text, imagery, and live or prerecorded video and audio, which he could call up by opening additional windows on the screen. Through the theater data base, he had access to State Department reports, Defense Intelligence Agency summaries, Central Intelligence Agency accounts, and National Imagery and Mapping Agency charts. Likewise, he could call up the latest tactical reports and analyses by a variety of categories—time, unit, contents, location, reliability—and could specify the level of information resolution— "granularity," they called it. Any time he asked for tactical reports over a period of time, the software would automatically "crunch out" a trend analysis, both in picture and bullet form. With a little manipulation, he could get direct feeds from satellites or aerial reconnaissance drones. (This procedure was not taught in the classroom; it was an unauthorized "back door" gateway, but nobody complained when Davis pulled it off.) Perhaps most important of all, he could access the Cable News Network for the latest-breaking developments. There was no lack of information out there, Davis knew. You were being bombarded by it. Any yahoo could access a near-endless flow of impressive data. The trick to being a good computer operator was being able to sift through it all to access the right information in the right form at the right time so the old man could figure out what it meant.

In an effort to make some sense of the enemy situation, they pulled down various "packages" of information, mostly in picture form, which promptly appeared and disappeared on the screen at Davis' command. Enemy armor spottings within the last 48, 24, and 12 hours. All ground contacts reported in the last 48 and 24 hours. All enemy artillery units spotted and fire missions reported in the last 48 hours. Road and rail usage in the last 72 hours. Sightings of enemy mobile air defense equipment, usually a good indicator of the disposition of the main body, in the last 48 and 24 hours. Enemy radio traffic in the last week. Enemy aviation activity in the last 2 days. Every once in a while the MEF commander would ask for a "template," a computer-generated estimate of possible enemy dispositions and movements based on the partial information that was available. Each template automatically came with a reliability estimate—"resolution," they called it—calculated as a percentage of complete reliability. The best resolution they had gotten for any one template was 45 percent; most were in the twenties and thirties. Statistically not very good—but certainly as good as could be expected.

Another set of red enemy symbols flashed on the screen.

"What the hell," the MEF commander said, looking at the

screen which indicated a heavy flow of enemy helicopter traffic along a single route. *A major heliborne operation? In this weather? As if things aren't sticky enough. And why is this the first I'm hearing of it?* "You're telling me the enemy's been flying fleets of helicopters continuously the last 6 hours?"

Vanderwood looked to MSgt Edgar Tomlinson, the intel- ligence chief.

"No, general," Tomlinson said. "He's not flying anything. What you're seeing on the screen, believe it or not, is actually a row of power lines. We checked it out. Radiating and blowing in this wind, our sensors picked them up as helicopters."

"You're kidding me, Top," the MEF commander said skeptically. "Our sensors think a set of power lines is a bunch of helicopters?"

"I guarantee it, general," Tomlinson said. "If you want to call up an aerial photo, I can show you the power lines."

"No, I believe you, Top."

"I've seen it happen before," Tomlinson said. "This gear is great, as long as you don't trick yourself into thinking that it's actually smart."

Despite an aggregate resolution of under 25 percent, Vanderwood sensed that a possible pattern had slowly begun to develop, but hardly anything conclusive. A possibility. A hunch. A little better than a wild guess. Despite the admittedly amazing technology, you could never be certain of anything, Vanderwood knew. Despite the artificial intelligence, the decision aids, the computer analysis. As long as war re- mained a clash of human wills, Vanderwood mused, no mat- ter how much technology you had, it still boiled down in the end to intuition and judgment.

"General, the division commander's coming in on video link," a Marine interrupted.

A window opened on the wall screen, and MajGen Miles Bishop appeared, apparently from inside a command AAV somewhere on the battlefield, the trademark cigar stub clamped in his teeth.

"Hey, can anybody hear me?" he was saying gruffly over the background noise in the AAV. "Is this blasted thing on?"

"Bish, this is Vanderwood with the MEF commander," the wing commander said. "You're coming in fine on this end."

"The video whatzit thing is on the blink on this end, but I can hear you okay," Bishop replied.

"Glad you could spare a few minutes out of your busy schedule," the MEF commander said. "We've been trying to figure out what the hell's been going on. We've been running some software for the last half hour, and we think we might have something."

"You want to know what the hell's going on, general?" Bishop said. "Hell, I can tell you what's going on."

"Okay, let us have it," the MEF commander said, and Bishop proceeded to describe in his own colorful but accurate way the same situation that had begun to take shape, with much less clarity, on the wall screen of the TACC. Vanderwood and the MEF commander exchanged glances. *Bishop*, Vanderwood mused, shaking his head. *What a piece of work. Glad he's on my team.*

"How did you come by that, Bish?" the MEF commander asked.

"Me and a couple of the boys sitting around a heat tab making some coffee just swagged it," Bishop said with a lopsided grin. "Ever-lovin' coop da oil—isn't that what you're always calling it, Harry?"

"*Coup d'oeil*, Bish," Vanderwood pronounced—referring to the French term which described the ability of gifted commanders to peer through the "fog of war" and intuitively grasp what was happening on the battlefield.

"Yeah, whatever," Bishop snorted.

Vanderwood grinned at Bishop's famous good-old-boy routine. Outside the circle of general officers, few Marines knew that French was one of the four foreign languages that Bishop spoke like a native.

"As long as he's got it," the MEF commander said, "let him pronounce it however he likes."

1428 Tuesday: Capt Takashima heard the unmistakable sound of the ATGMs firing off in unison like a naval broadside. The doctrinal manuals called it "massed, surprise fires." Takashima called it "a world of hurt for the bad guys." *Damn if those bastards didn't walk right into it*, he thought as he scampered forward to get a better look at the situation at the crossroads where first platoon had just sprung an ambush on the leading elements of the enemy column. *I owe Knutsen a beer when this is all over.* He couldn't explain how he knew, but just from the sound of things he could tell that first platoon

17

had caught them pretty good. Amazing how you learned to sense these things. The ground nearby erupted in a massive explosion, and he hit the deck—or rather, the 6 inches of water that covered the deck.

"Olsen, you all right?" he yelled after checking to make sure he was still in one piece.

"Yessir," his radioman replied. "Captain, third platoon wants to talk to you."

Second Lieutenant Tim Dandridge, Golf Company's least experienced platoon commander, was several hundred yards off to the right. Takashima had originally put third platoon where he could keep his eye on Dandridge, but when he'd spun the company, it had left third platoon off on the right flank by itself. Takashima switched on his headset.

"Oscar-3, this is Romeo-2-Oscar, go."

"Romeo-2-Oscar, I've got mechanized activity to my front and more activity moving through the woods around my right flank, over," Dandridge reported.

Even over the radio Takashima could sense the nervousness in the lieutenant's voice.

"Echo's on your right flank," Takashima said.

"Roger, Romeo-2-Oscar, I don't think it's Echo," came the reply. "I'm not picking them up on PLRS."

Takashima checked his electronic map board, networked to Olsen's radio, which in addition to his own eight-digit location could show the location of friendly transmitters. He punched in a request for the location of all transmitters of platoon level or higher. Dandridge was right: no Echo Company units. Which meant one of two things: either Echo was so badly lost they weren't even on the map, or somebody had keyed the wrong code into all of Echo's transmitters.

"Have you made contact with Echo?" Takashima asked.

"Negative. Can't raise them."

"Any visual with the enemy?"

"Negative, but they're definitely out there," Dandridge said. "Estimate at least a company."

"Roger, are you in position yet?" Third platoon should have been well set in by now, ready to ambush the advancing enemy forces.

There was a pause. "Er, roger . . . pretty much, Romeo-2-Oscar," came the halting reply.

Which meant "No," Takashima knew. Good news got passed without hesitation; bad news always seemed to move more reluctantly. Not a good sign. For a second, he considered heading over to third platoon's position to check things out, but he quickly dismissed the idea. His intuition still told him the critical action was taking place in front of him at first platoon's position. Events were still unfolding as expected, thanks to Knut-case. This was where he needed to be. Chances were that the young lieutenant was exaggerating; but yet, if Dandridge was right, then Takashima had read things wrong, and the enemy had other ideas in mind. You could never count on the bastards doing what they were supposed to.

"Gunny!" Takashima bellowed over the sound of the shelling.

A moment later GySgt Roberto Hernandez splashed down beside him.

"Gunny, third platoon is reporting enemy activity to their front and flank," Takashima began.

"Roger that, skipper," Hernandez said. "I was listening in."

Naturally, Takashima thought. Nothing the gunny did surprised him anymore.

"I'm concerned about what's going on over there," Takashima said. "But I don't have time to check it out myself. That activity they reported might or might not be Echo Company. Gunny, I want you to hustle over there, have a look around, and report back to me what you see. Use an alternate net. If it's real trouble, I need to know in a hurry. Don't step on any toes, but you might want to make a few tactful suggestions if it's appropriate."

"You want me to be, sir, what is sometimes referred to in the literature as a 'directed telescope,' " Hernandez said.

"Directed tele-what? Get outta here, gunny," Takashima said with a grin.

Sometimes it was a pain having the best-read staff NCO in the Marine Corps as a company gunny, he decided as he watched Hernandez charge away. But not usually.

1455, Tuesday: "Any questions?"

Any questions? Col Perry Gorman, the division G-3, wondered incredulously. *Where should I start?*

MajGen Bishop had just spent the last half hour orienting his staff to the new situation. He stood in front of the large electronic mapboard in the musty tent which housed the division's future ops section. The map was crisscrossed with the broad arrows and symbols he had been drawing with the stylus while he talked. Every once in a while Bishop would call for an estimate or opinion, or one of his staff would ask a question, make a recommendation, or take the stylus to sketch on the map. An energetic discussion would usually ensue and Bishop would let this go on for a few minutes, listening to the arguments for and against and benignly chewing on his cigar while the members of his staff had their say; then he would suddenly shut the discussion off and announce his position. Sometimes Bishop followed the advice of his staff; sometimes, Gorman was convinced, the general had already made his decision but wanted to make sure his people felt that they had had the opportunity to participate. It was truly an education watching Bishop work his staff, Gorman decided. It was a fluid and idiosyncratic process, reflective of Bishop's own personality. Never exactly the same twice and yet very effective. Anybody who thought staff planning was a mechanical process had never been around MajGen Miles Bishop.

There was an old military saying, attributed to the Prussian Field Marshal Moltke, that no plan survives contact with the enemy. In a short period of time by merely modifying an existing branch plan, Bishop quickly reoriented the efforts of the division to meet the new situation. Gorman's first thought was for the wasted effort; but he quickly realized the effort had not been wasted at all: it had been a valuable learning process which had resulted in an improved situational awareness that was shared by Bishop, the entire staff, and subordinate commanders.

A feeling had engulfed the command post that through pre-vious good planning and adaptability the division had turned a potential crisis into a decisive opportunity. Of course, Gorman mused, an awful lot of things had to happen to make adaptabil-ity during execution possible. *It's amazing how much prepara-tion is required to provide flexibility in execution.* A division contained an awful lot of independent parts that needed to be working toward the same goal. The intelligence collection plan would have to be reoriented to the new axis of advance, as would the fire support planning and the logistics effort. Poten-tial enemy countermoves would have to be considered, as well as possible ways to deal with them. *One good thing that does-n't have to change is the commander's intent and its end state.* The force would have to be reorganized to support the new taskings. Fragmentary orders would have to be issued. Necessary coordination would have to be effected above, be-low, and laterally—especially with the wing since all the avia-tion support requirements had changed. The light armor battalion would have to be redeployed to continue the counter-reconnaissance battle. With Task Force Hammer as well as all the forward units committed to the exploitation, a new reserve would have to be constituted somehow, but not immediately. Thought would have to be given to protecting the lengthening lines of communications as the pursuit continued. The gener-al's concept for a regimental helicopterborne attack into the en-emy rear would have to be worked out—a major evolution in itself (although most of the planning and coordination would be done by the regiment). Landing zones and helicopter lanes would have to be reconnoitered, air defenses located and targeted for suppression . . .

"Last chance," Bishop was saying. "No saved rounds?"

"You want this by when exactly, general?" a voice from the back of the tent asked.

Laughter broke out, and Bishop smiled but did not bother to answer. The general's obsession with tempo was legendary.

"Look, people, don't worry about trying to control every moving piece in this monster. It's not gonna happen. I can tell you it's gonna be chaos for the next few days at least. Maybe longer. The battalions and regiments are already starting to do what we need them to do, so let's not try to overcontol this thing. I just want you to make sure that all the chaos and mayhem are flowing in the same general direction and that we keep it going. Coordinate what absolutely needs to be coordinated and don't try to coordinate what doesn't. Keep this thing pointed straight, but let it go. Remember, the sign of a good plan is that it gives you both direction and flexibility.

"All right," the general concluded, "I think everybody knows where we stand and what needs to be done. Let's get at it."

1505 Tuesday: If 2ndLt Connors had been unhappy before, he was positively miserable now. He decided he felt about as useful as a mindless pawn in some giant chess game, being moved around one square at a time. *Certainly don't want to get too far ahead of ourselves, do we?* The analogy was pretty appropriate, he thought. Too bad the chess player who was ordering him around showed every sign of being an indecisive beginner who seemed to be taking an awful lot of time between moves.

What made things worse was that from the distant shelling and the radio traffic he could tell that there was one heck of a battle going on. And he was missing it. Every time he radioed for instructions he'd get the same reply—"Wait out"—and when the orders eventually arrived, it seemed that he was always one step too late. Usually, he'd arrive just in time to have to duck the tail end of somebody else's fire mission. He might just as well have been wandering around the pine forests of beloved Camp Lejeune for all the action he was getting. Was that a red-cockaded woodpecker he just saw?

He crawled to the edge of the vegetation and peered across the clearing. That was the objective, all right, some 300 yards away. Hill 124, now known as Objective Rose after the company commander's mother-in-law. He checked his watch; the prep fires were scheduled to commence at half past. He searched the hilltop carefully through his thermal binoculars and saw no sign whatsoever of enemy activity.

Of course, he didn't know what he had expected to see. He'd been given no information on the enemy situation on the objective, and he had no idea why he was attacking this hill in the first place. He certainly had no idea what made Hill 124 so important—other than that it was a convenient place to draw a goose egg on some higher-up's map. He was expected to attack and seize Objective Rose, commencing at 1530, and that was that.

With his squad leaders, Connors crawled back to rejoin the platoon.

"Lieutenant, company wants to talk to you," his radioman

reported.

Connors switched on his headset.

"Hotel-3-Mike, this is 3-Mike-2," he said.

"3-Mike-2, are you ready yet?"

For about the tenth time, Connors said to himself. *Keep your shorts on; the attack doesn't go for another 20 minutes.*

"Roger that," he replied out loud. "The objective is deserted."

"Roger. The prep fires commence at 1530, as scheduled, and last for 5 mikes."

"I say again, the objective is deserted, Hotel-3-Mike," Connors said. "We don't need the prep fires; we can just walk up on the objective."

There was a pause. Connors could imagine the captain wrestling with that one. He couldn't blame the captain, really. Battalion wanted things done a certain way. To change things now, Connors knew, would throw off the timetable and would mean shutting off the scheduled fires—in short, it would disrupt the plan. And you certainly didn't want to interfere with the plan, he knew—not in this battalion anyway. The plan was everything. All the elements of the battalion were supposed to attack in close synchronization, Connors knew— "synchronization" was LtCol Hewson's favorite buzzword. Of course, when Connors thought of synchronization he invariably thought of synchronized swimming, and he smiled at the ridiculous image of a couple of swimmers pirouetting in graceful unison in a pool. He couldn't imagine anything less like combat than that. *I might not have a world of experience,* he thought, *but how could anybody in his right mind think you could*

25

synchronize the confusion and mayhem of any military operation? It boggled the mind.

"Listen, the prep starts at 1530," came the eventual reply over the radio, somewhat testily. "Just do it."

"Roger, out," Connors said resignedly. *Three bags full.*

Setting in the base of fire and getting the other two squads in position for the assault was the work of only a few minutes. Connors checked his watch: still only a quarter past. *Hope nobody falls asleep waiting*, he thought, abundantly aware of Marines' remarkable ability to doze off on a moment's notice anytime, anyplace, in any conditions. Fifteen minutes later, exactly on schedule, the preparation fires commenced and ended 5 minutes later. Battalion would be pleased: the attack went flawlessly; there was no enemy resistance to screw it up. His two squads swept through the tall grass toward the hill and within minutes were consolidating on the objective. There was no sign that the enemy had ever occupied the hill. Whether the enemy was anywhere in the vicinity he couldn't tell: because of the tall grass, visibility was about 10 yards in any direction. No matter; they had accomplished the assigned task.

"Lieutenant, company gunny wants our ammo and casualty report," his radioman said.

Connors chuckled scornfully. *Seeing as there was nobody to shoot at us and nobody for us to shoot at . . . . Now, Connors, you're being a malcontent again. Just go through the motions and don't make waves.*

"I'll take it," he said, switching on his headset. "Hotel-3-Mike, this is 3-Mike-2. No casualties; no ammo expended. Mission accomplished. What next, over?"

There was pause, then finally the reply came: "Wait out."

1635 Tuesday: "Romeo-2-Oscar, this is 2-Oscar-3."

Capt Takashima recognized the gunny's voice on the command alternate net. He had been hearing the firefight coming from that direction for some time now—he didn't know how long: it could be 20 minutes, it could be 2 hours—but he didn't have time to think about it. He'd feel a lot better once he got the gunny's opinion of the situation.

"This is 2-Oscar, go," Takashima said.

"Confirm situation as described earlier by 2-Oscar-3," Hernandez reported. "Engaged, situation well in hand. Echo was a little slow getting their act together, but Oscar-3 saved their butts. Caught the enemy pretty good."

"Have you made contact with Echo?"

"Roger," Hernandez said. "Have been attached."

"Say again," Takashima said, confused.

"2-Oscar-3 has been attached to Alpha-7-Hotel."

What the hell? Who the hell does Schuler think he is, taking it on himself to attach one of my platoons to his company? Takashima was about to cut loose with some choice words, but he thought better of it. He knew that he was in no posi- tion to try to control what third platoon was doing; he was too busy dealing with the situation at the crossroads. Sometimes the enemy didn't use the same boundaries that we did, Takashima realized: third platoon was really part of

Echo's fight. That being the case, Takashima knew that for the purposes of unity of command third platoon ought to be answering to Schuler and not to him. It was hardly conventional, Takashima decided—certainly not the school solution—but, under the circumstances, it was the right thing to do. *I guess that's what gunny would call a "self-organizing, complex adaptive system,"* Takashima mused. *I'll just have to remember to give Schuler a hard time about needing four platoons to do what we can do with only two.*

0255 Wednesday: The MEF commander shed his dripping poncho as he stepped out of the rain into the MEF command post. The military policeman snapped to attention and saluted.

"Carry on, Sgt McDavid. Cpl Cooper," he said to his soaked driver, "get some sleep. It's been a long day."

He made his way into the operations center and dropped wearily into his chair where he'd started the operation some 24 hours before. In the last 24 hours, he'd been all over the MEF area of operations. He'd been to the division forward command post to talk to the division commander face-to-face about how to deal with the unexpected developments. He'd insisted on a face-to-face because he wanted to make sure they understood each other. He'd been to the wing head- quarters twice to try to get a handle on the overall situation and to see what could be done about air support. He'd personally debriefed the Cobra pilots who'd first spotted the enemy column. He'd videotaped a new intent statement—an "intent-o-mmercial," as the Marines jokingly referred to it—to be broadcast to the entire MEF (at least down to battalion and squadron level, the lowest level that had video capability). He and the wing commander had taken a terrifying V-22 flight over the battlefield (and unfortunately had gotten precious little out of it). He'd been back to the MEF command post once during the day to see if the situation had gotten any clearer since he'd left: it hadn't. He'd visited the division's main-effort regiment and that regiment's main-effort battalion near Culverin Crossroads. (It hadn't been until he'd met that CO from Golf Company, Capt Taka-something, and had seen the indomitable fighting spirit of his Marines that he'd realized that the MEF would carry the day—"Just get me some air, general," the captain had said.) He'd visited the engineers to make sure that the roads were going to hold up for at least the next 72 hours in this rain. He'd even spent several hours supervising an assault river crossing during the critical early stages of the pursuit. And he'd happened upon the FSSG commander at a maintenance contact point, of all places, where they'd watched an M1A3 main battle tank repaired and put back into action; they'd discussed the logistics needed to support the upcoming exploitation.

It seemed like days since he'd been at the command post. On the wall screen before him, the amorphous wave of flashing green unit symbols had crept considerably farther north since the last time he had looked at the map. There were far more red enemy symbols now as well, most of them encircled in the lower left-hand part of the screen, an indication that the intelligence effort had managed to locate many of the enemy forces

29

that had been unknown at the beginning of the operation. He knew that many of those units, although still reflected on the map and still present on the ground, had ceased to be effective fighting forces by now. He also knew that the clean image portrayed on the screen could not begin to capture the brutal fighting and the destruction that he had witnessed during the day. That was the great danger of being stuck in a command post, he knew; you began to confuse what was on a map with reality.

Based on the tempo of activity in the operations center, he wondered if the staff knew that the battle was all but won. In the next room, the major and the two staff sergeants who made up the future plans cell would be working feverishly on the plans for the next week to exploit the advantage the MEF had won today. *The responsibility of command is never finished,* he decided; *always something else to be done.* Curiously, he thought, he found himself thinking back to his days as a brand-new lieutenant at The Basic School, remembering the adage that had been drilled into them: "Camouflage is continuous." *Command is continuous*, he found himself thinking. *I'll have to remember that one,* he decided, *for the next time I'm invited to speak at a TBS mess night.*

He thought of stopping in to see how things were going in the future plans cell, but he knew his chief of staff would have things moving along briskly, and he would just be getting in the way. Even when the issue had still hung in the balance, Col Dick Westerby had been pushing the future ops guys to develop a plan for exploiting the outcome. Around the MEF command element, Westerby was known with a certain grudging admiration as "Yesterday," because that was when he seemed to want everything done. "If it's not done fast," Westerby was fond of saying, "it's not done right."

As if by cue, the small, balding colonel appeared, bearing a cup of steaming coffee.

"You look like you could use this, general," he said.

"Thanks, Dick," the MEF commander said.

A staff sergeant appeared. "Here's the new MEF op order, colonel," he said, handing a flimsy document to Westerby.

Westerby perused the two-page order which consisted of a page of text and a diagram, nodding as he read.

"General, do you want to have a look at this?" he asked.

"Hell, no. I couldn't even focus my eyes on it. That's what I've got you for. You know my intent."

"Looks good, Staff Sergeant Walters," Westerby said, initialing the document and returning it to the staff sergeant. "Let's get it out 10 minutes ago."

"Aye, aye, sir; it'll go straight out on the secure fax," Walters said and quickly departed.

The MEF commander sipped his coffee and gazed at the large situation screen.

"Well, what do you think, Dick?" he asked.

"What do I think?" Westerby said. "I think we went in with an unclear picture of the situation, and it only got worse. As is usually the case, the enemy tended not to cooperate. Weather precluded using the bulk of our aviation and restricted the mobility of some of our vehicles. Our original plan had to be quickly discarded and another put in its place. We had to adapt to a rapidly changing situation. Our previous planning efforts provided us with the flexibility and situational awareness to react to a changing situation and provided flexibility to our subordinates. Thank goodness for staff officers, pilots, and subordinate commanders who exercise initiative and quickly adapt to changing situations."

"Yes," the general said with obvious satisfaction, "don't you love it when the system works to perfection?"

Chapter 1

The Nature
of
Command and Control

"War is the realm of uncertainty; three quarters of the factors on which action in war is based are wrapped in a fog of greater or lesser uncertainty. . . . The commander must work in a medium which his eyes cannot see; which his best deductive powers cannot always fathom; and with which, because of constant changes, he can rarely become familiar."

—Carl von Clausewitz

To put effective command and control into practice, we must first understand its fundamental nature—its purpose, characteristics, environment, and basic functioning. This understanding will become the basis for developing a theory and a practical philosophy of command and control.

HOW IMPORTANT IS COMMAND AND CONTROL?

No single activity in war is more important than command and control. Command and control by itself will not drive home a single attack against an enemy force. It will not destroy a single enemy target. It will not effect a single emergency resupply. Yet none of these essential warfighting activities, or any others, would be possible without effective command and control. Without command and control, cam- paigns, battles, and organized engagements are impossible, military units degenerate into mobs, and the subordination of military force to policy is replaced by random violence. In short, command and control is essential to all military operations and activities.

With command and control, the countless activities a military force must perform gain purpose and direction. Done well, command and control adds to our strength. Done poorly, it invites disaster, even against a weaker enemy. Command and control helps commanders make the most of what they have—people, information, material, and, often most important of all, time.

In the broadest sense, command and control applies far beyond military forces and military operations. Any system comprising multiple, interacting elements, from societies to sports teams to any living organism, needs some form of command and control. Simply put, command and control in some form or another is essential to survival and success in any competitive or cooperative enterprise. Command and control is a fundamental requirement for life and growth, survival, and success for any system.

WHAT IS COMMAND AND CONTROL?

We often think of command and control as a distinct and specialized function—like logistics, intelligence, electronic warfare, or administration—with its own peculiar methods, considerations, and vocabulary, and occurring independently of other functions. But in fact, command and control encompasses all military functions and operations, giving them meaning and harmonizing them into a meaningful whole. None of the above functions, or any others, would be purposeful without command and control. Command and control is not the business of specialists—unless we consider the commander a specialist—because command and control is fundamentally the business of the commander.[1]

Command and control is the means by which a commander recognizes what needs to be done and sees to it that appropriate actions are taken. Sometimes this recognition takes the form of a conscious command decision—as in deciding on a concept of operations. Sometimes it takes the form of a preconditioned reaction—as in immediate-action drills, practiced in advance so that we can execute them reflexively in a moment of crisis. Sometimes it takes the form of a rules-based procedure—as in the guiding of an aircraft on final approach. Some types of command and control must occur so quickly and precisely that they can be accomplished only by computers—such as the command and control of a guided missile in flight. Other forms may require such a degree of judgment and intuition that they can be performed only by skilled, experienced people—as in devising tactics, operations, and strategies.

Sometimes command and control occurs concurrently with the action being undertaken—in the form of real-time guidance or direction in response to a changing situation. Sometimes it occurs beforehand and even after. Planning, whether rapid/time-sensitive or deliberate, which determines aims and objectives, develops concepts of operations, allocates resources, and provides for necessary coordination, is an important element of command and control. Furthermore, planning increases knowledge and elevates situational awareness.

Effective training and education, which make it more likely that subordinates will take the proper action in combat, establish command and control before the fact. The immediate-action drill mentioned earlier, practiced beforehand, provides command and control. A commander's intent, expressed clearly before the evolution begins, is an essential part of command and control. Likewise, analysis after the fact, which ascertains the results and lessons of the action and so informs future actions, contributes to command and control.

Some forms of command and control are primarily procedural or technical in nature—such as the control of air traffic and air space, the coordination of supporting arms, or the fire control of a weapons system. Others deal with the overall conduct of military actions, whether on a large or small scale, and involve formulating concepts, deploying forces, allocating resources, supervising, and so on. This last form of command and control, the overall conduct of military actions, is our primary concern in this manual. Unless otherwise specified, it is to this form that we refer.

Since war is a conflict between opposing wills, we can measure the effectiveness of command and control only in relation to the enemy. As a practical matter, therefore, effective command and control involves protecting our own command and control activities against enemy interference and actively monitoring, manipulating, and disrupting the enemy's command and control activities.

WHAT IS THE BASIS OF COMMAND AND CONTROL?

The basis for all command and control is the authority vested in a commander over subordinates. Authority derives from two sources. Official authority is a function of rank and position and is bestowed by organization and by law. Personal authority is a function of personal influence and derives from factors such as experience, reputation, skill, character, and personal example. It is bestowed by the other members of the organization. Official authority provides the power to act but is rarely enough; most effective commanders also possess a high degree of personal authority. Responsibility, or accountability for results, is a natural corollary of authority. Where there is authority, there must be responsibility in like measure. Conversely, where individuals have responsibility for achieving results, they must also have the authority to initiate the necessary actions.[2]

WHAT IS THE RELATIONSHIP BETWEEN "COMMAND" AND "CONTROL"?

The traditional view of command and control sees "command" and "control" as operating in the same direction: from the top of the organization toward the bottom.[3] (See figure 1.) Commanders impose control on those under their command; commanders are "in control" of their subordinates, and subordinates are "under the control" of their commanders.

We suggest a different and more dynamic view of command and control which sees command as the exercise of authority and control as feedback about the effects of the ac- tion taken. (See figure 1.) The commander commands by deciding what needs to be done and by directing or influencing the conduct of others. Control takes the form of feedback—the continuous flow of information about the unfolding situation returning to the commander—which allows the commander to adjust and modify command action as needed. Feedback indicates the dif- ference between the goals and the situation as it exists. Feed- back may come from any direction and in any form—intelligence about how the enemy is reacting, informa- tion about the status of subordinate or adjacent units, or re- vised guidance from above based on developments. Feedback is the mechanism that allows commanders to adapt to changing circumstances—to exploit fleeting opportunities, respond to developing problems, modify schemes, or redirect efforts. In this way, feedback "controls" subsequent command action. In such a command and control system, control is not strictly something that seniors impose on subordinates; rather, the en- tire system comes "under control" based on feedback about the changing situation.[4]

Command and control is thus an interactive process involv- ing all the parts of the system and working in all directions. The result is a mutually supporting system of give

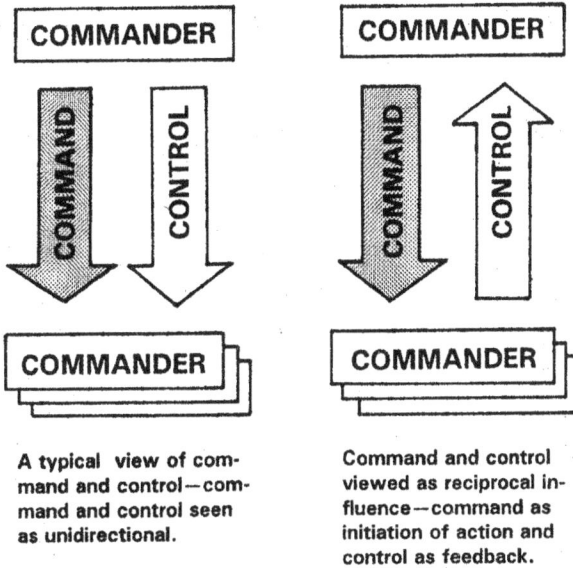

A typical view of command and control—command and control seen as unidirectional.

Command and control viewed as reciprocal influence—command as initiation of action and control as feedback.

Figure 1. Two views of the relationship between command and control.

and take in which complementary commanding and controlling forces interact to ensure that the force as a whole can adapt continuously to changing requirements.

WHAT DOES IT MEAN TO BE "IN CONTROL"?

The typical understanding of effective command and control is that someone "in command" should also be "in control." Typically, we think of a strong, coercive type of command and control—a sort of pushbutton control—by which those "in control" dictate the actions of others and those "under control" respond promptly and precisely, as a chess player controls the movements of the chess pieces. But given the nature of war, can commanders control their forces with anything even resembling the omnipotence of the chess player? We might say that a gunner is in control of a weapon system or that a pilot is in control of an aircraft. But is a flight leader really directly in control of how the other pilots fly their aircraft? Is a senior commander really in control of the squads of Marines actually engaging the enemy, especially on a modern battlefield on which units and individuals will often be widely dispersed, even to the point of isolation?

We are also fond of saying that commanders should be "in control" of the situation or that the situation is "under control." The worst thing that can happen to a commander is to "lose" control of the situation. But are the terrain and weather under the commander's control? Are commanders even remotely in control of what the enemy does? Good commanders may sometimes anticipate the enemy's actions and may even influence the enemy's actions by seizing the initiative and forcing the enemy to react to them. But it is a delusion to believe that we can truly be in control of the enemy or the situation.[5]

The truth is that, given the nature of war, it is a delusion to think that we can be in control with any sort of certitude or precision. And the further removed commanders are from the Marines actually engaging the enemy, the less direct control they have over their actions. We must keep in mind that war is at base a human endeavor. In war, unlike in chess, "pieces" consist of human beings, all reacting to the situation as it pertains to each one separately, each trying to survive, each prone to making mistakes, and each subject to the vagaries of human nature. We could not get people to act like mindless robots, even if we wanted to.

Given the nature of war, the remarkable thing is not that commanders cannot be thoroughly in control but rather that they can achieve much influence at all. We should accept that the proper object of command and control is not to be thoroughly and precisely in control. The turbulence of modern war suggests a need for a looser form of influence—some- thing that is more akin to the willing cooperation of a basketball team than to the omnipotent direction of the chess player—that provides the necessary guidance in an uncertain, disorderly, time-competitive environment without stifling the initiative of subordinates.

COMPLEXITY IN COMMAND AND CONTROL

Military organizations and military evolutions are complex systems. War is an even more complex phenomenon—our complex system interacting with the enemy's complex system in a fiercely competitive way. A complex system is any system composed of multiple parts, each of which must act individually according to its own circumstances and which, by so acting, changes the circumstances affecting all the other parts. A boxer bobbing and weaving and trading punches with his opponent is a complex system. A soccer team is a complex system, as is the other team, as is the competitive interaction between them. A squad-sized combat patrol, changing formation as it moves across the terrain and reacting to the enemy situation, is a complex system. A battle between two military forces is itself a complex system.[6]

Each individual part of a complex system may itself be a complex system—as in the military, in which a company consists of several platoons and a platoon comprises several squads—creating multiple levels of complexity. But even if this is not so, even if each of the parts is fairly simple in itself, the result of the *interactions* among the parts is highly complicated, unpredictable, and even uncontrollable behavior. Each part often affects other parts in ways that simply cannot be anticipated, and it is from these unpredictable interactions that complexity emerges. With a complex system it is usually extremely difficult, if not impossible, to isolate individual causes and their effects since the parts are all connected in a complex web. The behavior of complex systems is frequently nonlinear which means that even extremely small influences can have decisively large effects, or vice versa. Clausewitz wrote that "success is not due simply to general causes. Particular factors can often be decisive—details only known to those who were on the spot . . . while issues can be decided by chances and incidents so minute as to figure in histories simply as anecdotes." [7] The element of chance, interacting randomly with the various parts of the system, introduces even more complexity and unpredictability.

It is not simply the number of parts that makes a system complex: it is the way those parts interact. A machine can be complicated and consist of numerous parts, but the parts generally interact in a specific, designed way—or else the machine will not function. While some systems behave mechanistically, complex systems most definitely do not. Complex systems tend to be open systems, interacting frequently and freely with other systems and the external environment. Complex systems tend to behave more "organically"—that is, more like biological organisms.[8]

The fundamental point is that any military action, by its very nature a complex system, will exhibit messy, unpredictable, and often chaotic behavior that defies orderly, efficient, and precise control. Our approach to command and control must find a way to cope with this inherent complexity. While a machine operator may be in control of the machine, it is difficult to imagine any commander being in control of a complex phenomenon like war.

This view of command and control as a complex system characterized by reciprocal action and feedback has several important features which distinguish it from the typical view of command and control and which are central to our approach. First, this view recognizes that effective command and control must be sensitive to changes in the situation. This view sees the military organization as an open system, interacting with its surroundings (especially the enemy), rather than as a closed system focused on internal efficiency. An effective command and control system provides the means to adapt to changing conditions. We can thus look at command and control as a process of continuous adaptation. We might better liken the military organization to a predatory animal—seeking information, learning, and adapting in its quest for survival and success—than to some "lean, green machine." Like a living organism, a military organization is never in a state of stable equilibrium but is instead in a continuous state of flux—continuously adjusting to its sur- roundings.

Second, the action-feedback loop makes command and control a continuous, cyclic process and not a sequence of discrete actions—as we will discuss in greater detail later. Third, the action-feedback loop also makes command and control a dynamic, interactive process of cooperation. As we have discussed, command and control is not so much a matter of one part of the organization "getting control over" another as something that connects all the elements together in a cooperative effort. All parts of the organization contribute action and feedback—"command" and "control"—in overall cooperation. Command and control is thus fundamentally an activity of *reciprocal influence*—give and take among all parts, from top to bottom and side to side.

Fourth, as a result, this view does not see the commander as being above the system, exerting command and control from the outside—like a chess player moving the chess pieces—but as being an integral part of this complex web of reciprocal influence. And finally, as we have mentioned, this view recognizes that it is unreasonable to expect command and control to provide precise, predictable, and mechanistic order to a complex undertaking like war.

WHAT MAKES UP COMMAND AND CONTROL?

The words "command" and "control" can be nouns,[9] and used in this way the phrase *command and control* describes a system—an arrangement of different elements that interact to produce effective and harmonious actions. The basic elements of our command and control system are people, information, and the command and control support structure.

The first element of command and control is *people*—people who gather information, make decisions, take action, communicate, and cooperate with one another in the accomplishment of a common goal. People drive the command and control system—they make things happen—and the rest of the system exists only to serve them. The essence of war is a clash between human wills, and any concept of command and control must recognize this first. Because of this human element, command is inseparable from leadership. The aim of command and control is not to eliminate or lessen the role of people or to make people act like robots, but rather to help them perform better. Human beings—from the senior commander framing a strategic concept to a lance corporal calling in a situation report—are integral components of the command and control system and not merely users of it.

All Marines feel the effects of fear, privation, and fatigue. Each has unique, intangible qualities which cannot be captured by any organizational chart, procedure, or piece of equipment. The human mind has a capacity for judgment, intuition, and imagination far superior to the analytical capacity of even the most powerful computer. It is precisely this aspect of the human element that makes war in general, and command in particular, ultimately an art rather than a science. An effective command and control system must account for the characteristics and limits of human nature and at the same time exploit and enhance uniquely human skills. At any level, the key individual in the command and control system is the commander who has the final responsibility for success.

The second element of command and control is *information*, which refers to representations of reality which we use to "inform"—to give form and character to—our decisions and actions. Information is the words, letters, numbers, images, and symbols we use to represent things, events, ideas, and values. In one way or another, command and control is essentially about information: getting it, judging its value, processing it into useful form, acting on it, sharing it with others. Information is how we give structure and shape to the material world, and it thus allows us to give meaning to and to gain understanding of the events and conditions which surround us. In a very broad sense, information is a control parameter: it allows us to provide control or structure to our actions.[10]

The value of information exists in time since information most often describes fleeting conditions. Most information grows stale with time, valuable one moment but irrelevant or even misleading the next.

There are two basic uses for information. The first is to help create situational awareness as the basis for a decision. The second is to direct and coordinate actions in the execution of the decision. While distinct in concept, the two uses of information are rarely mutually exclusive in practice. There will usually be quite a bit of overlap since the same exchange of information often serves both purposes simultaneously. For example, coordination between adjacent units as they execute the plan can also help shape each unit's understanding of the situation and so inform future decisions. An order issued to subordinates describes the tasks to be accomplished and provides necessary coordinating instructions; but the same order should provide a subordinate insight into the larger situation and into how the subordinate's actions fit into that larger situation. Likewise, a call for fire, the primary purpose of which is to request supporting arms from a supporting unit, also provides information about the developing situation in the form of a target location and description.

Information forms range from data—raw, unprocessed signals—to information that has been evaluated and integrated into meaningful knowledge and understanding. A commander's guidance to the staff and orders to subordinates constitute information as do intelligence about the enemy, status reports from subordinate units, or coordination between adjacent units. Without the information that provides the basis of situational awareness, no commander—no matter how experienced or wise—can make sound decisions. Without information that conveys understanding of the concept and intent, subordinates cannot act properly. Without information in the form of a strike brief which provides understanding of the situation on the ground, a pilot cannot provide close air support. Without information which provides understanding of an upcoming operation and the status of supply, the logistician cannot provide adequate combat service support.

Effective command and control is not simply a matter of generating enough information. Most information is not important or even relevant. Much is unusable given the time available. More is inaccurate, and some can actually be misleading. Given information-gathering capabilities today, there is the distinct danger of overwhelming commanders with more information than they can possibly assimilate. In other words, too much information is as bad as too little—and probably just as likely to occur. Some kinds of information can be counterproductive—information which misleads us, which spreads panic, or which leads to overcontrol. Information is valuable only insofar as it contributes to effective decisions and actions. *The critical thing is not the amount of information, but key elements of information, available when needed and in a useful form, which improve the commander's awareness of the situation and ability to act.*

The final element of command and control is the *command and control support structure*[11] which aids the people who create, disseminate, and use information. It includes the organizations, procedures, equipment, facilities, training, education, and doctrine which support command and control. It is important to note that although we often refer to families of hardware as "systems" themselves, the command and control system is much more than simply equipment. High-quality equipment and advanced technology do not guarantee effective command and control. Effective command and control starts with qualified people and an effective guiding philosophy. We must recognize that the components of the command and control support structure do not exist for their own sake but solely to help people recognize what needs to be done and take the appropriate action.

WHAT DOES COMMAND AND CONTROL DO?

The words "command" and "control" are also verbs,[12] and used that way, the phrase *command and control* describes a process—a collection of related activities. We draw an important distinction between a process, a collection of related activities, and a procedure, a specific sequence of steps for accomplishing a specific task. Command and control is a process. It may include procedures for performing certain tasks, but it is not itself a procedure and should not be approached as one.

Command and control is something we *do*. These activities include, but are not limited to, gathering and analyzing information, making decisions, organizing resources, planning, communicating instructions and other information, coordinating, monitoring results, and supervising execution.

As we seek to improve command and control, we should not become so wrapped up in feeding and perfecting the process that we lose sight of the object of command and control in the first place. For example, we should not become so con- cerned with the ability to gather and analyze huge amounts of information efficiently that we lose sight of the primary goal of helping the commander gain a true awareness of the situation as the basis for making and implementing decisions. The ultimate object is not an efficient command and control process; the ultimate objective is the effective conduct of military action.

So rather than ask what are the functions that make up command and control, we might better ask: What should effective command and control do for us? First, it should help provide insight into the nature and requirements of the problem facing us. It should help develop intelligence about the enemy and the surroundings. As much as possible, it should help to identify enemy capabilities, intentions, and vulnerabilities. It should help us understand our own situation—to include identifying our own vulnerabilities. In short, it should help us gain situational awareness.

Next, command and control should help us devise suit-able and meaningful goals and adapt those goals as the situation changes. It should help us devise appropriate actions to achieve those goals. It should help us provide direction and focus to create vigorous and harmonious action among the various elements of the force. It should help us provide a means of continuously monitoring developments as the basis for adapting. It should provide security to deny the enemy knowledge of our true intentions. And above all, it should help generate tempo of action since we recognize that speed is a weapon.

So, what does command and control do? In short, effective command and control helps generate swift, appropriate, decisive, harmonious, and secure action.

THE ENVIRONMENT OF COMMAND AND CONTROL: UNCERTAINTY AND TIME

The defining problem of command and control that overwhelms all others is the need to deal with uncertainty.[13] Were it not for uncertainty, command and control would be a simple matter of managing resources. In the words of Carl von Clausewitz, "War is the realm of uncertainty; three quarters of the factors on which action in war is based are wrapped in a fog of greater or lesser uncertainty. A sensitive and discriminating judgment is called for; a skilled intelligence to scent out the truth."[14]

Uncertainty is what we do not know about a given situation—which is usually a great deal, even in the best of circumstances. We can think of uncertainty as doubt which blocks or threatens to block action.[15] Uncertainty pervades the battlefield in the form of unknowns about the enemy, about the surroundings, and even about our own forces. We may be uncertain about existing conditions—factual information— such as the location and strength of enemy forces. But even if we are reasonably sure about factual information, we will be less certain of what to infer from those facts. What are the enemy's intentions, for example? And even if we make a reasonable inference from the available facts, we cannot know which of the countless possible eventualities will occur.

In short, uncertainty is a fundamental attribute of war. We strive to reduce uncertainty to a manageable level by gathering and using information, but we must accept that we can never eliminate it. Why is this so? First, since war is fundamentally a human enterprise, it is shaped by human nature and is subject to the complexities, inconsistencies, and peculiarities which characterize human behavior. Human beings, friendly as well as enemy, are unpredictable. Second, because war is a complex struggle between *independent* human wills, we can never expect to anticipate with certainty what events will develop. In other words, the fundamentally complex and interactive nature of war *generates* uncertainty. Uncertainty is not merely an existing environmental condition; it is a natural byproduct of war.

Command and control aims to reduce the amount of un- certainty that commanders must deal with—to a reasonable point—so they can make sound decisions. Though we try to reduce uncertainty by providing information, there will always be some knowledge that we lack. We will be aware of some of the gaps in our knowledge, but we will not even be aware of other unknowns. We must understand the forces that guarantee uncertainty and resolve to act despite it on the basis of what we do know.

It is important to note that certainty is a function of knowledge and understanding and not merely of data. Although they are clearly related—they are all forms of information, as we will discuss—the distinctions among them are important. Data serve as the raw material for knowledge and understanding. *Knowledge and understanding result when human beings add meaning to data.* Properly provided and processed, data can lead to knowledge and understanding, but the terms are not synonymous. Paradoxically, not all data lead to knowledge and understanding; some may even hamper the gaining of knowledge and understanding. The essential lesson from this distinction is that decreased uncertainty is not simply a matter of increased information flow. More important are the quality of the information and the abilities of the person using it—and the willingness and ability to make decisions in the face of uncertainty.

The second main element that affects command and control, second only to uncertainty in order of importance, is the factor of time. Theoretically, we can always reduce uncertainty by gaining more knowledge of the situation (accepting that there is some information we can never gain). The basic dilemma is that to gain and process information takes time. This creates three related problems. First, the knowledge we gain in war is perishable: as we take the time to gain new information, information already gained is becoming obsolete. Second, since war is a contest between opposing wills, time itself is a precious commodity used by both sides. While we strive to get information about a particular situation, the enemy may already be acting—and changing the situation in the process. (Of course, the enemy faces the same problem in relation to us.) And third, the rapid tempo of modern operations limits the amount of information that can be gathered, processed, and assimilated in time to be of use. Command and control thus becomes a tense race against time. So the second absolute requirement in any command and control system is to be fast—at least faster than the enemy.

The resulting tension between coping with uncertainty and racing against time presents the fundamental challenge of command and control. This is perhaps the single most important point to take from this chapter. It is also important to recognize that the enemy faces the same problems—and the object is to achieve some relative advantage. Although there is no easy answer to this problem, the successful commander must find a solution, as we will discuss.

COMMAND AND CONTROL IN THE INFORMATION AGE

Many of the factors that influence command and control are timeless—the nature of war and of human beings and the twin problems of uncertainty and time, for example. On the other hand, numerous factors are peculiar to a particular age or at least dependent on the characteristics of that age. As war has evolved through the ages, so has command and control. In general, as war has become increasingly complicated, so have the means of command and control. What can we conclude about the environment in which command and control must function today and in the foreseeable future?

The prevailing characteristics of the information age are variety and rapid, ongoing change. An unstable and changeable world situation can lead to countless varieties of conflict requiring peacekeeping operations on the one extreme to general war on the other. Since we cannot predict when and where the next crisis will arise or what form it will take, our command and control must function effectively in any envi- ronment.

Technological improvements in mobility, range, lethality, and information-gathering continue to compress time and space, forcing higher operating tempos and creating a greater demand for information. Military forces may move more quickly over greater distances than ever before, engaging the enemy at greater ranges than ever before. The consequence of this is fluid, rapidly changing military situations. The more quickly the situation changes, the greater the need for continuously updated information and the greater the strain on command and control. Future conflict will require military forces able to adapt quickly to a variety of unexpected circumstances.

The increasing lethality and range of weapons over time has compelled military forces to disperse in order to survive, similarly stretching the limits of command and control. Military forces are bigger and more complex than ever before, consisting of a greater number and variety of specialized organizations and weapons. As a result, modern military forces require ever greater amounts of information in order to operate and sustain themselves, even in a peacetime routine.

In the current age, technology is increasingly important to command and control. Advances in technology provide capabilities never before dreamed of. But technology is not without its dangers, namely the overreliance on equipment on the one hand and the failure to fully exploit the latest capabilities on the other. It is tempting, but a mistake, to believe that technology will solve all the problems of command and control. Many hopes of a decisive technological leap forward have been dashed by unexpected complications and side effects or by the inevitable rise of effective countermeasures. Moreover, used unwisely, technology can be part of the problem, contributing to information overload and feeding the dangerous illusion that certainty and precision in war are not only desirable, but attainable.

In this complicated age, command and control is espe- cially vulnerable and not just to the physical destruction of facilities and personnel by enemy attack. As the command and control system becomes increasingly complex, it likewise becomes increasingly vulnerable to disruption, monitoring, and penetration by the enemy as well as to the negative side effects of its own complicated functioning. Its own complexity can make command and control vulnerable to disruption by information overload, the overreliance on technology, misinformation, communications interference, lack of human understanding, lack of technical proficiency or training, mechanical breakdown, and systemic failure.

CONCLUSION

Although command and control systems have evolved continuously throughout history, the fundamental nature of command in war is timeless. Noteworthy improvements in technology, organization, and procedures have not eased the demands of command and control at all and probably never will. While these improvements have increased the span of command and control, they have barely kept pace with the increasing dispersion of forces and complexity of war itself. Whatever the age or technology, the key to effective command and control will come down to dealing with the fundamental problems of uncertainty and time. Whatever the age or technology, effective command and control will come down to people using information to decide and act wisely. And whatever the age or technology, the ultimate measure of command and control effectiveness will always be the same: Can it help us act faster and more effectively than the enemy?

Chapter 2

Command and Control Theory

"Confronted with a task, and having less information available than is needed to perform that task, an organization may react in either of two ways. One is to increase its information-processing capacity, the other to design the organization, and indeed the task itself, in such a way as to enable it to operate on the basis of less information. These approaches are exhaustive; no others are conceivable. A failure to adopt one or the other will automatically result in a drop in the level of performance."

—Martin van Creveld, *Command in War*

H aving reached a common understanding of the nature of command and control, we turn to developing a theory about the command and control process that will in turn serve as the basis for creating an effective command and control system.

POINT OF DEPARTURE: THE OODA LOOP

Our study of command and control theory starts with a simple model of the command and control process known as the OODA loop.[1] The OODA loop applies to any two-sided conflict, whether the antagonists are individuals in hand-to-hand combat or large military formations. OODA is an acronym for observation-orientation-decision-action, which describes the basic sequence of the command and control process. (See figure 2.) When engaged in conflict, we first observe the situation—that is, we take in information about our own status, our surroundings, and our enemy. Sometimes we actively seek the information; sometimes it is thrust upon us. Having observed the situation, we next orient to it—we make certain estimates, assumptions, analyses, and judgments about the situation in order to create a cohesive mental image. In other words, we try to figure out what the situation means to us. Based on our orientation, we decide what to do—whether that decision takes the form of an immediate reaction or a de- liberate plan. Then we put the decision into action. This in-

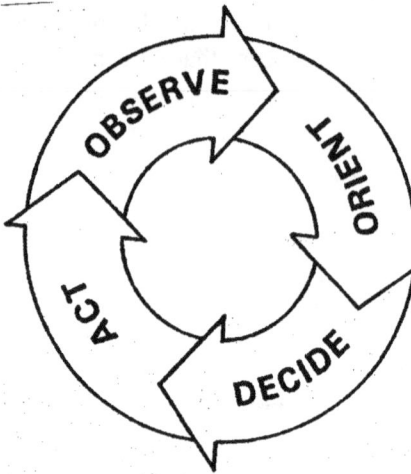

**Figure 2. The command and control process:
The OODA loop.**

cludes disseminating the decision, supervising to ensure proper
execution, and monitoring results through feedback, which
takes us full circle to the observation phase. Having acted, we
have changed the situation, and so the cycle begins again. It is
worth noting that, in any organization with multiple decision
makers, multiple OODA loops spin simultaneously, although
not necessarily at the same speed, as commanders exercise
command and control at their own level and locale.

Importantly, the OODA loop reflects how command and control is a continuous, cyclical process. In any conflict, the antagonist who can consistently and effectively cycle through the OODA loop faster—who can maintain a higher tempo of actions—gains an ever-increasing advantage with each cycle. With each reaction, the slower antagonist falls farther and farther behind and becomes increasingly unable to cope with the deteriorating situation. With each cycle, the slower antagonist's actions become less relevant to the true situation. Command and control itself deteriorates.

The lesson of the OODA loop is the importance of generating tempo in command and control. In other words, *speed is an essential element of effective command and control*. Speed in command and control means shortening the time needed to make decisions, plan, coordinate, and communicate. Since war is competitive, it is not absolute speed that matters, but speed relative to the enemy: the aim is to be faster than our enemy, which means interfering with the enemy's command and control as well as streamlining our own. The speed differential does not necessarily have to be a large one: a small advantage exploited repeatedly can quickly lead to decisive results. We should recognize that the ability and desire to generate a higher operational tempo does not negate the willingness to bide time when the situation calls for patience. The aim is not merely rapid action, but also meaningful action.

THE INFORMATION HIERARCHY

We use the term *information* generically to refer to all manner of descriptions or representations from raw signals on the one hand to knowledge and understanding on the other. But it is important to recognize that there are actually four different classes of information. We must understand the differences between these classes because they are of different value in supporting command and control. (See figure 3.) We must also understand what happens to information as it moves between levels on the hierarchy.[2]

Raw data comprise the lowest class of information and include raw signals picked up by a sensor of any kind (a radio antenna, an eyeball, a radar, a satellite) or communicated between any kind of nodes in a system. Data are bits and bytes transferred between computers, individual transmissions sent by telephone or radio or facsimile, or a piece of unprocessed film. In other words, raw data are signals which have not been processed, correlated, integrated, evaluated, or interpreted in any way. This class of information is rarely of much use until transformed in some way to give it some sort of meaning.

The next class is data that have been processed into or have been displayed in a form that is understandable to the people

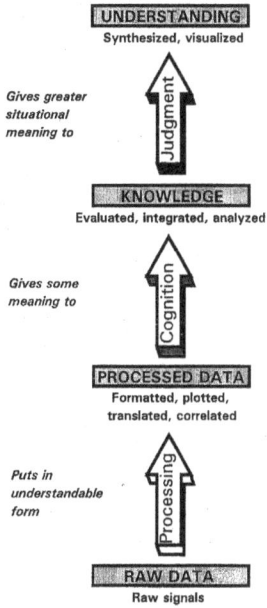

Figure 3. The information hierarchy.

who must use them.* Processed data include film that has been developed into a photograph, radio transmissions copied into a standard report format, a computer file displayed as text or a graphic on a screen, grid coordinates plotted on a map, or an intercepted enemy message deciphered. The act of processing in itself gives the data a limited amount of value. Clearly, processed data are more useful to people than raw data—and some may have immediate, obvious and signifi-cant value—but they have not yet been evaluated or analyzed.

The next rung on the information hierarchy is knowledge—data that have been analyzed to provide meaning and value. Knowledge is data which have been evaluated as to reliability, relevance, and importance. Knowledge is various pieces of processed data which have been integrated and interpreted to begin to build a picture of the situation. For example, military intelligence is a form of knowledge as compared to combat information which has not yet undergone analysis and evaluation. Likewise, situation reports pieced together to create an estimate of the situation represent knowledge. At this level, we are starting to get a product which can be useful for decisionmaking.

---

* This class is often referred to as "information," in a more specific usage of that term. To avoid confusion, we will continue to refer to this class as "processed data" and will use "information" to refer to the full range of information classes.

The highest class of information is understanding—knowledge that has been synthesized and applied to a specific situation to gain a deeper level of awareness of that situation. We may *know* what is going on; we *understand* why. Understanding results when we synthesize bodies of knowledge, use judgment and intuition to fill in the gaps, and arrive at a complete mental image of the situation. Understanding means we have gained situational awareness. Understanding reveals the critical factors in any situation. It reveals the enemy's critical vulnerabilities. It reveals the patterns and logic of a situation. Understanding thus allows us to anticipate events—to recognize in advance the consequences of new or impending developments or the effects of our actions on the enemy. We try to make understanding the basis for our decisions—although recognizing that we will rarely be able to gain full under- standing.

The gradations between the different classes of information are not always very clear. It is not always easy to tell the exact difference between raw and processed data, for example. But it is important to realize that there are differences and that knowledge is usually more valuable than data, for instance. Moreover, it is also important to recognize that information is transformed as it moves up the hierarchy and to understand the forces that cause that transformation.

Raw data are turned into processed data, as we might expect, through *processing*, an activity involving essentially the rote application of procedure. Processing includes formatting, translating, collating, plotting, and so on. Much processing occurs automatically (whether by humans or by machines) without our even being aware that it is taking place—such as when a facsimile machine converts bits of data into understandable text or graphics. In many cases, machines can process data much more quickly and efficiently than people.

We turn processed data into knowledge through the activity of *cognition*—the act of learning what something means, at least in general terms. To a degree, cognition may be based on rules of logic or deduction ("If *A* happens, it means *B*"). Expert systems and artificial intelligence can assist with cognition to a certain extent—by helping to integrate pieces of processed data, for example. But cognition is primarily a human mental activity—not primarily a procedural act like processing, but an act of learning.

We transform the complex components of knowledge into understanding through *judgment*, a purely human skill based on experience and intuition, beyond the capability of any current artificial intelligence or expert system. Judgment simply cannot be reduced to procedures or rules (no matter how complex).

We should note that as information moves up the hierarchy from data toward understanding, an integration occurs. Multiple bits of raw data are pieced together to make processed data. Numerous pieces of processed data coalesce into knowledge. Various bodies of knowledge distill into understanding. This integration is essential to eventually reaching understanding because it involves reducing the total number of "pieces" that must be considered at any one time. The vast number of bits of raw data that describe any situation would overwhelm any commander *if they had to be considered singly*. It takes a certain amount of time and effort to make these integrations, but without this effort the commander would be overloaded by a staggering number of things to consider.

By nature, data are significantly easier to generate, identify, quantify, reproduce, and transmit than are knowledge and understanding. But commanders need knowledge and understanding in order to make effective decisions. Likewise, subordinates need not merely data but knowledge and understanding of the commander's concept and intent. The goal in command and control should not be collecting, processing, and communicating vast amounts of data—and increasing the danger of information overload in the process—but approaching understanding as closely as possible. However, we cannot simply provide commanders with ready-made understanding. They will have to make the final judgments themselves. But we can strive to provide information that is as easily assimilable and as close to final form as possible. This means providing information in the form of images.

IMAGE THEORY

Human beings do not normally think in terms of data or even knowledge. People generally think in terms of ideas or images—mental pictures of a given situation. Not only do people generally think in images, they understand things best as images and are inspired most by images.[3]

We can say that an image is the embodiment of our understanding of a given situation or condition. (The term *coup d'oeil*, which refers to the ability of gifted commanders to intuitively grasp what is happening on the battlefield, means literally "stroke of the eye.") Images apply not only to the military problems we face but also to the solutions. For example, a well-conceived concept of operations and commander's intent should convey a clear and powerful image of the action and the desired outcome.

People assimilate information more quickly and effectively as visual images than in text. The implications of this are widespread and significant, ranging from technical matters of presentation—the use of maps, overlays, symbols, pictures, and other graphics to display and convey information visually—to conceptual matters of sharing situational awareness and intent.

Our image of a situation is based not just on the facts of the situation, but also on our interpretation of those facts. In other words, it is based on our intuition, appreciation, judgment, and so on, which in turn are the products of our preconceptions, training, and past experiences. New information that does not agree with our existing image requires us to revalidate the image or revise it—not easily done in the turbulence and stress of combat. The images we create and communicate to others must approximate reality. Conversely, if we want to deceive our enemies, we try to present them with an image of the situation that does not match reality and so lead them to make poor decisions.

We generate images from others' observations as well as our own. In general, the higher the level of command, the more we depend on information from others and the less on our own observations. All but the smallest-unit commanders receive most of their information from others. This can cause several problems. First, when we observe a situation firsthand, we have an intuitive appreciation for the level of uncertainty—we have a sense for how reliable the image is—and we can act accordingly. But when we receive our information secondhand, we usually lose that sense. This is especially dangerous in a high-technology age in which impressively displayed information appears especially reliable. Second, we can sense more about a situation from firsthand observation than we can faithfully communicate to others or, at least, than we have time to communicate in a crisis. Third, since each of us interprets events differently, the information we communicate is distorted to some degree with each node that it passes through on its way to its final destination. And fourth, this same information is likewise delayed at each node. Since the value of information exists in time, this delay can be critical.

Commanders need essentially three different pictures. The first is a closeup of the situation, a "feel" for the action gained best through personal observation and experience. From this picture, commanders gain a sense of what subordinates are experiencing—their physical and moral state. From this image, commanders get a sense of what they can and cannot demand of their people. In the words of Israeli General Yshayahor Gavish about his experience in the 1967 Arab-Israeli war: "There is no alternative to looking into a subordinate's eyes, listening to his tone of voice." [4]

The second picture is an overall view of the situation. From this view, commanders try to make sense of the relative dispositions of forces and the overall patterns of the unfolding situation. From this view, they also gauge the difference between the actual situation and the desired end state. The desired result of the overall view is a quality we can call "topsight"—a grasp of the big picture. If "*insight* is the illumination to be achieved by penetrating inner depths, *topsight* is what comes from a far-overhead vantage point, from a bird's eye view that reveals *the whole*—the big picture; how the parts fit together." [5]

The third picture we try to form is the action as seen through the eyes of the enemy commander from which we try to deduce possible enemy intentions and anticipate possible enemy moves. Of the three pictures, the first is clearly the most detailed but usually offers a very narrow field of vision. Commanders who focus only on this image risk losing sight of the big picture. The second picture provides an overall image but lacks critical detail—just as a situation map does not capture more than a broad impression of the reality of events on the battlefield. Commanders who focus only on this image risk being out of touch with reality. The third picture is largely a mental exercise limited by the fact that we can never be sure of what our enemy is up to.

Squad leaders or fighter pilots may simultaneously be able to generate all three images largely from their own observations. Higher commanders, however, feel a tension between satisfying the need for both the closeup and overall images—the former best satisfied by personal observation at the front and the latter probably best satisfied from a more distant vantage point, such as a command post or higher headquarters.

As we have mentioned, any system which attempts to communicate information by transmitting images will suffer from a certain degree of distortion and delay. There are several ways to deal with this problem. The first is for commanders to view critical events directly to the greatest extent possible (consistent with the competing need to stay abreast of the overall situation). In this way they avoid the distortions and delays which occur when information filters through successive echelons.

Because as war has evolved, it has become increasingly complex and dispersed, commanders have found it increasingly difficult to observe all, or even most, critical events directly. One historical solution to this problem is a technique known as the *directed telescope*, which can be especially useful for gaining a closeup image. This technique involves using a dedicated information collector—whether a trusted and like-minded subordinate or a sensor—to observe selected events and report directly to the commander. Commanders may direct the "telescope" at the enemy, at the surroundings, or at their own forces. In theory, because these observers report directly, the information arrives with minimal delay or distortion. Directed telescopes should not replace regular reporting chains but should augment them—to avoid burdening lower echelons with additional information gathering and to check the validity of information flowing through regular channels. Improperly used, directed telescopes can damage the vital trust a commander seeks to build with subordinates.[6]

The second way to deal with the problems of delay and distortion of information is to rely on *implicit communications* to the greatest extent possible. Implicit communication minimizes the need for explicit transmission of information. Theoretically, because implicit communication requires indi- viduals who share a common perspective, information will suffer minimal distortion as it passes up or down the chain. We will discuss implicit communication in greater detail later.

The third way to deal with the problems of delay and distortion of information, also discussed later in more detail, is to decentralize decisionmaking authority so that the individual on the spot, the individual who has direct observation of the situation at that spot, is the person making the decisions.

THE COMMAND AND CONTROL SPECTRUM

Historically, there have been two basic responses to the fundamental problem of uncertainty: to pursue certainty as the basis for effective command and control or to accept uncertainty as a fact and to learn to function in spite of it.

The first response to uncertainty is to try to minimize it by creating a powerful, highly efficient command and control apparatus able to process huge amounts of information and intended to reduce nearly all unknowns. The result is *detailed* command and control. Such a system stems from the belief that if we can impose order and certainty on the disorderly and uncertain battlefield, then successful results are predictable. Such a system tends to be technology-intensive.

Detailed command and control can be described as *coercive*, a term which effectively describes the manner by which the commander achieves unity of effort.[7] In such a system, the commander holds a tight rein, commanding by personal direction or detailed directive.[8] Command and control tends to be centralized and formal. Orders and plans are detailed and explicit, and their successful execution requires strict obedience and minimizes subordinate decisionmaking and initiative. Detailed command and control emphasizes vertical, linear information flow: in general, information flows up the chain of command and orders flow down. Discipline and coordination are imposed from above to ensure compliance with the plan.

In a system based on detailed command and control, the command and control process tends to move slowly: information must be fed up to the top of the chain where sole decisionmaking authority resides, and orders must filter to the bottom to be executed. Understandably, such a system does not generally react well to rapidly changing situations. Nor does it function well when the vertical flow of information is disrupted. While distrust is not an inherent feature of detailed command and control, organizations characterized by distrust tend toward detailed command and control.

This approach represents an attempt to overcome the fundamental nature of war. Since we have already concluded that precise direction is generally impossible in war, detailed command and control risks falling short of its desired result. The question is whether it nears the desired result enough to achieve overall success.

By contrast, *mission* command and control accepts the turbulence and uncertainty of war. Rather than increase the level of certainty that we seek, by mission command and control we reduce the degree of certainty that we need. Mission command and control can be described as *spontaneous*: unity of effort is not the product of conformity imposed from above but of the spontaneous cooperation of all the elements of the force.[9] Subordinates are guided not by detailed instructions and control measures but by their knowledge of the require- ments of the overall mission. In such a system, the command- er holds a loose rein, allowing subordinates significant freedom of action and requiring them to act with initiative. Discipline imposed from above is reinforced with self-discipline throughout the organization. Because it decentralizes decisionmaking authority and grants subordinates significant freedom of action, mission command and control demands more of leaders at all levels and requires rigorous training and edu- cation.

Mission command and control tends to be decentralized, informal, and flexible. Orders and plans are as brief and simple as possible, relying on subordinates to effect the necessary coordination and on the human capacity for implicit communication—mutual understanding with minimal information exchange. By decentralizing decisionmaking authority, mission command and control seeks to increase tempo and improve the ability to deal with fluid and disorderly situations.

Moreover, with its reliance on implicit communications, mission command and control is less vulnerable to disruption of the information flow than is detailed command and control.

The two approaches to the problem mark the theoretical extremes of a spectrum of command and control. (See figure 4.) In practice, no commander will rely entirely on either purely detailed or purely mission methods. Exactly what type of command and control we use in a particular situation will depend on a variety of factors, such as the nature of the action or task, the nature and capabilities of the enemy, and, perhaps most of all, the qualities of our people. This is not to suggest that the two types of command and control are of equal value and merely a matter of personal preference. While detailed command and control may be appropriate in the performance of specific tasks of a procedural or technical nature, it is less than effective in the overall conduct of military operations in an environment of uncertainty, friction, disorder, and fleeting opportunities, in which judgment, creativity, and initiative are required. Militaries have frequently favored detailed command and control, but our understanding of the true nature of war and the lessons of history points to the advantages of mission command and control.

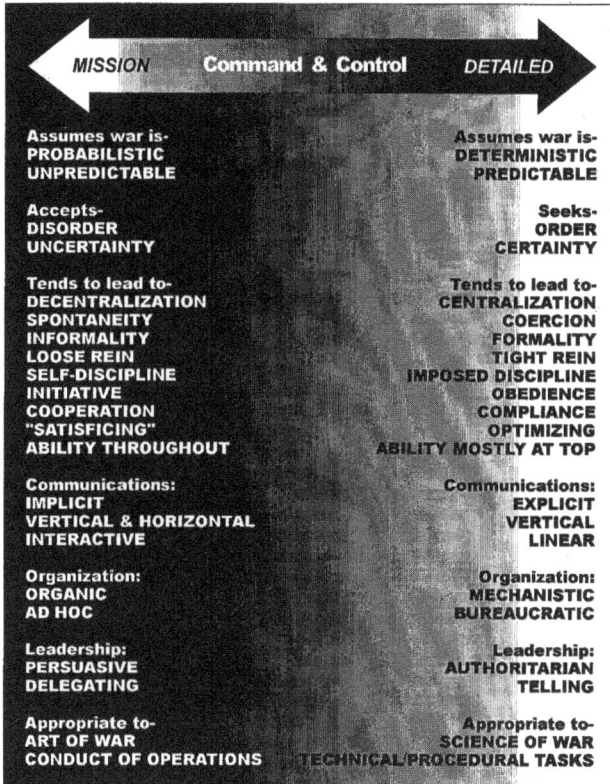

MISSION   Command & Control   DETAILED

Assumes war is-	Assumes war is-
PROBABILISTIC	DETERMINISTIC
UNPREDICTABLE	PREDICTABLE
Accepts-	Seeks-
DISORDER	ORDER
UNCERTAINTY	CERTAINTY
Tends to lead to-	Tends to lead to-
DECENTRALIZATION	CENTRALIZATION
SPONTANEITY	COERCION
INFORMALITY	FORMALITY
LOOSE REIN	TIGHT REIN
SELF-DISCIPLINE	IMPOSED DISCIPLINE
INITIATIVE	OBEDIENCE
COOPERATION	COMPLIANCE
"SATISFICING"	OPTIMIZING
ABILITY THROUGHOUT	ABILITY MOSTLY AT TOP
Communications:	Communications:
IMPLICIT	EXPLICIT
VERTICAL & HORIZONTAL	VERTICAL
INTERACTIVE	LINEAR
Organization:	Organization:
ORGANIC	MECHANISTIC
AD HOC	BUREAUCRATIC
Leadership:	Leadership:
PERSUASIVE	AUTHORITARIAN
DELEGATING	TELLING
Appropriate to-	Appropriate to-
ART OF WAR	SCIENCE OF WAR
CONDUCT OF OPERATIONS	TECHNICAL/PROCEDURAL TASKS

Figure 4. The Command and control spectrum.

LEADERSHIP THEORY

Leadership is the influencing of people to work toward the accomplishment of a common objective. Because war is fundamentally a human endeavor, leadership is essential to effective command and control. There are two basic theories of leadership that generally correspond to the theories of command and control.

The authoritarian theory of leadership is based on the assumption that people naturally dislike work and will try to avoid it where possible, and that they must therefore be forced by coercion and threat of punishment to work toward the common goal. This theory further argues that people actually prefer to be directed and try to avoid responsibility. The result is an autocratic style of leadership aimed at achieving immediate and unquestioning obedience. Leaders announce their decisions and expect subordinates to execute them. The authoritarian leader is sometimes also known as a *telling* or *directing* leader. While authoritarian leadership may result in rapid obedience, it also can often result in subordinates who are highly dependent on the leader, require continuous supervision, and lack initiative. Military discipline is widely seen as an example of this model since quick and unquestioning response to orders may be required in the heat of an emergency. This is, however, only one version of leadership that military leaders have used successfully.

The opposite theory of leadership, known as persuasive or delegating leadership, assumes that work is as natural as rest or play, that people do not inherently dislike work, and that work can be either a source of satisfaction (in which case people will perform it willingly) or a source of punishment (in which case they will avoid it). This theory rejects the idea that external supervision and the threat of punishment are the most effective ways to get people to work toward the common objective. The persuasive theory argues that people will exercise initiative and self-control to the degree they are committed to the organizational objective. Under proper conditions, people learn not only to accept responsibility but to actively seek it. According to this theory, the potential for exercising imagination, ingenuity, and creativity in the solution of unit problems is widespread throughout any unit. Leadership thus becomes a question of inspiring, guiding, and supporting committed subordinates and encouraging them to perform freely within set limits. Over time, delegating or persuasive leadership tends to produce subordinates who exhibit a high degree of independence, self-discipline, and initia- tive.[10]

The leadership style we adopt in a given situation depends on a variety of factors. Key among them is the maturity of subordinates—that is, how motivated, experienced, and willing to accept responsibility they are. Here maturity is not necessarily linked to age or seniority. The more mature the subordinate, the more we can delegate; the less mature, the more we will have to direct. All other things being equal, we prefer the persuasive approach because it seeks to gain the committed performance of subordinates and encourages subordinate initiative.

Moreover, persuasive leadership reduces the need for continuous supervision, an important consideration on a dispersed and fluid battlefield on which continuous, detailed supervision is problematic.

PLANNING THEORY

Planning is the process of developing practical schemes for taking future actions. Planning may occur before a decision and so support decisionmaking—by analyzing the mission, the enemy, or the environment to help develop situational awareness or by studying the feasibility of different courses of action. Planning may also occur after a decision and so support its execution—by working out necessary coordination measures, allocation of resources, or timing and scheduling.

Planning facilitates future decisions and actions by helping commanders provide for those things which are not likely to change or which are fairly predictable (such as geography and certain aspects of supply or transport). Planning helps them to examine their assumptions, to come to a common understanding about the situation and its general direction, to anticipate possible enemy actions, and thus to consider possible counteractions. Planning helps to uncover and clarify potential opportunities and threats and to prepare for opportunities and threats in advance. Conversely, planning helps to avoid preventable mistakes and missed opportunities.

By definition, planning is oriented on the future. It represents an effort to project our thoughts and designs forward in time and space. Because the future is always uncertain, planning should generally not seek to specify future actions with precision. The farther ahead we plan, the more time we allow ourselves to prepare, but the less certain and specific our plans can be. Planning ahead thus becomes less a matter of trying to direct events and more a matter of identifying options and possibilities.

Depending on the situation and the nature of the preparations, planning may be done rapidly or deliberately. Rapid/time-sensitive planning is conducted in response to existing conditions and is meant for immediate or near-future execution. In contrast, deliberate planning is based on anticipated future conditions and is intended for possible execution at some more distant time. We should keep in mind that all planning takes time and must facilitate the generation or maintenance of tempo, while ensuring that time allocated for planning does not adversely impact on tempo.

Planning routines can improve the proficiency of a staff by creating an effectiveness and efficiency of effort. The goal of the Marine Corps is to develop an institutionalized planning framework for use at all echelons of command. However, we must guard against using an institutionalized planning framework in a lock-step fashion. We must ensure that the application of this planning process contributes to flexibility in conducting operations.

Planning occurs at different levels and manifests itself differently at these levels. At the highest level is what we can call conceptual planning which establishes aims, objectives, and intents and which involves developing tactical, opera- tional, or strategic concepts for the overall conduct of military actions. Conceptual planning should provide the foundation for all subsequent planning, which we can call functional and detailed. These are the more routine and pragmatic elements of planning which are concerned with translating the concept into a complete and practicable plan. Functional planning is concerned with the various functional areas necessary to sup- port the overall concept, such as subordinate concepts for mobilization, deployment, logistics, intelligence, and so on. Detailed planning encompasses the practical specifics of execution. Detailed planning deals primarily with scheduling, coordination, or technical matters required to move and sustain military forces, such as calculating the supplies or transport needed for a given operation.

In general, conceptual planning corresponds to the art of war, detailed planning applies to the science of war, and functional planning falls somewhere in between. Detailed and, to a lesser extent, functional planning may require deliberate and detailed calculations and may involve the development of detailed schedules or plans, such as landing tables, resupply schedules, communications plans, or task organizations. However, the staff procedures which may be necessary in de- tailed and functional matters are generally not appropriate for broader conceptual planning. Rather, such planning should attempt to broadly influence rather than precisely direct future actions. Conceptual planning should impart intent, develop overall operating concepts, and identify contingencies and possible problems but otherwise should leave the subordinate broad latitude in the manner of functional or detailed execution.

ORGANIZATION THEORY

Organization is an important tool of command and control. How we organize can complicate or simplify the problems of execution. By task-organizing our force into capable subor- dinate elements and assigning each its own task, we also organize the overall mission into manageable parts. The organization of our force, then, should reflect the conceptual organization of the plan.

Specifically, what should organization accomplish for us? First, organization establishes the chain of command and the command and support relationships within the force. The chain of command establishes authority and responsibility in an unbroken succession directly from one commander to an- other.

The commander at each level responds to orders and directions received from a higher commander and, in turn, issues orders and gives directions to subordinates. In this way, the chain of command fixes authority and responsibility at each level while at the same time distributing them broadly throughout the force; each commander has designated authority and responsibility in a given sphere. Command and support relationships specify the type and degree of authority one commander has over another and the type and degree of support that one commander provides another.

Importantly, organization should establish unity of command which means that any given mission falls within the authority and responsibility of a single commander and that a commander receives orders from only one superior for any given mission. Similarly, organization should ensure that a commander has authority over or access to all the resources required to accomplish the assigned mission.

Organization also serves the important socializing function of providing sources of group identity for members of the organization. For example, Marines may see themselves first as members of a squad, next as members of a platoon, and then as members of a company. An organization operates most effectively when its members think of themselves as belonging to one or more groups characterized by high levels of loyalty, cooperation, morale, and commitment to the group mission.

Each commander (supported by the staff) and immediate subordinates constitute an integrated team—a cohesive group committed to the accomplishment of a single mission. For example, a company commander and platoon commanders constitute a team cooperating in the accomplishment of the company mission. A platoon commander and squad leaders also constitute a team cooperating in the accomplishment of the platoon mission. The size of a team can vary with the situation, as we will discuss. Whereas the chain of command conveys authority and responsibility from commander to commander, the idea of an integrated team is to pull individuals together into cohesive groups. (See figure 5.) Each team functions as a single, self-contained organism—characterized by cooperation, reciprocal influence, lateral and vertical communication, and action-feedback loops operating continuously in all directions. Each member of the team may perform a different task, but always within the context of the team mission. Continuity throughout the organization results from each commander's being a member of two related teams, one as the senior and one as a subordinate.[11]

Organization should also provide commanders with staffs appropriate to the level of command. The staff assists the commander by providing specialized expertise and allowing a division of labor and a distribution of information. The staff is not part of the chain of command and thus has no formal authority in its own right, although commanders may delegate authority to a staff officer if they choose.

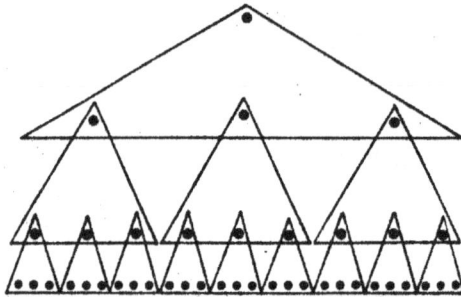

$\triangle$ = An integrated team, a cohesive group consisting of a commander (and by association his staff) and his immediate subordinates who work together toward the accomplishment of a common goal. The size of each team will depend on circumstances. All teams shown here in triangles are four persons.

Continuity is provided by the fact that each commander is a member of two related teams, one as the senior and one as a subordinate. Through this overlapping structure, the commander is able to extend his command over the entire force.

Figure 5. Overlapping units and teams.

Organization should ensure a reasonable span of control which refers to the number of subordinates or activities under a single commander. The span of control should not exceed a commander's capability to command effectively. The optimal number of subordinates is situation-dependent. For example, the more fluid and faster-changing a situation is, the fewer subordinate elements a commander can keep track of continuously. Likewise, commanders exercising detailed command and control, which requires them to pay close attention to the operations of each subordinate element, generally have narrower spans of control than commanders who use mission command and control and let their subordinates work out the details of execution.

Although a reasonable span of control varies with the situation, as a rule of thumb an individual can effectively command at least three and as many as seven subordinates. Within this situation-dependent range, a greater number means greater flexibility—three subordinate units allow for more options and combinations than two, for example. However, as the number increases, at some point we lose the ability to effectively consider each unit individually and begin to think of the units together as a single, inflexible mass. At this point, the only way to reintroduce flexibility is to group elements together into a smaller number of parts, thereby creating the need for another intermediate echelon of command. The evolution of the Marine rifle squad during the Second World War is a good example of this. Entering the war, the rifle squad consisted of nine Marines—a squad leader and eight squad members with no additional internal organization. In combat this squad lacked the flexibility needed for small-unit fire and maneuver. Moreover, squad leaders often could not effectively command eight individual Marines. The answer was the creation of an intermediate organizational level, the fire team of four Marines, which also allowed an increase in squad size to thirteen Marines. The creation of the fire team decreased the number of immediate subordinates the squad leader had to deal with, while extending the squad leader's influence over a larger squad.

Narrowing span of control—that is, lessening the number of immediate subordinates—means deepening the organization by adding layers of command. But the more layers of command an organization has, the longer it takes for information to move up or down. Consequently, the organization becomes slower and less responsive. Conversely, an effort to increase tempo by eliminating echelons of command, or flattening an organization, necessitates widening the span of control. The commander will have to resolve the resulting tension that exists between organizational width and depth. (See figure 6.)

Finally, organization does not apply only to people and equipment. It also applies to information. In large part, organization determines how we distribute information throughout the force and establishes communication channels.

Information may flow vertically within the chain of command, but it should not be restricted by the chain of command. It also flows laterally between adjacent units, or even "diagonally"—between a platoon and an adjacent company headquarters, for example, or between a supported unit and a supporting unit outside the chain of command. Information

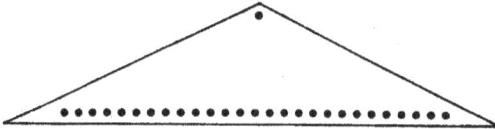

Example of an extremely "flat" organization with "wide" span of control; direct link between top and bottom.

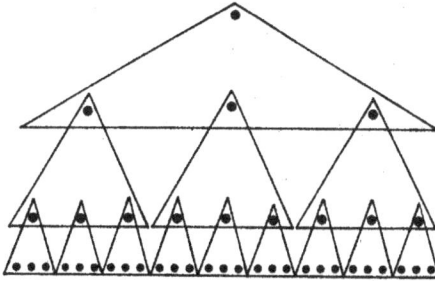

Example of the same unit organized "deeper" with "narrower" spans of control; never more than three "wide"; two intermediate echelons between top and bottom.

Figure 6. Organization width and depth.

flows informally and unofficially—that is, between individuals according to personal relationships—as well as according to formally established channels. These informal channels provide an important redundancy and are especially important in team building.

COMMUNICATIONS THEORY

Because military evolutions require cooperative effort, it is important that we be able to communicate effectively with others. Communications are any method or means of conveying information from one person or place to another to improve understanding. In general, effective organizations are characterized by intense, unconstrained communications— that is, the free and enthusiastic sharing of meaningful information throughout the organization.[12] Moreover, communication has an importance far beyond the exchange of information; it serves a socializing function. Separate from the quality or meaning of the information exchanged, the act of communicating strengthens bonds within an organization and so is an important device in building trust, cooperation, cohesion, and mutual understanding.

The traditional view of communications within military organizations is that the subordinate supplies the commander with information about the situation, and the commander in turn supplies the subordinate with decisions and instructions. This linear form of communication may be consistent with the exercise of detailed command and control, but not with a system based on mission command and control which instead requires interactive communications characterized by continuous feedback loops. Feedback provides the means to improve and confirm mutual understanding—and this applies to lateral as well as vertical communications.

We communicate by a wide variety of means: face-to-face conversation, radio, telephone, data link, written word, visual signal, picture, or diagram. Human beings communicate not only in the words they use, but also by tone of voice, inflection, facial expression, body language, and gestures. In fact, evidence suggests that in face-to-face conversation, humans actually communicate most by visual means (such as gestures, body language, or facial expressions), second by vocal non-verbal means (such as tone or inflection), and least by the actual words they use.[13]

Moreover, people can communicate implicitly—that is, they achieve mutual understanding and cooperation with a minimal amount of information having to be transmitted—if they have a familiarity formed of shared experiences and a common outlook. A key phrase or a slight gesture can sometimes communicate more than a detailed order. Since it reduces the time spent drafting and relaying messages, implicit communication also reduces the problems of delay typically associated with information flow. Implicit communication helps to maximize information content while minimizing the actual flow of data, thereby making the organization less vulnerable to the disruption of communications.

While conciseness is a virtue, so is a certain amount of redundancy. Used within reason, redundancy of communications can improve clarity of meaning and mitigate against disruptions to the communications system. Effective communications consequently exhibit a balance between conciseness and redundancy. (In general, the greater the implicit understanding within the organization, the less the need for redun- dancy.)

INFORMATION MANAGEMENT THEORY

Since effective command and control is concerned with getting the right information to the right person at the right time, information management is crucial.

We initiate communications under two basic principles: *supply-push* and *demand-pull*.[14] A supply-push system pushes information from the source to the user either as the information becomes available or according to a schedule. (See figure 7.) The advantages of supply-push are that the commander does not need to request the information and that the information generally arrives in a timely fashion. The challenge with a supply-push system is to be able to anticipate the commander's information needs. The danger of information overload arises primarily from supply-push.

By contrast, a pure demand-pull system does not rely on the ability to anticipate information needs; it is inactive until a demand is made on it. In a pure demand-pull system, the user generates all information requirements. (See figure 7.) If the information is readily available—already resident in some

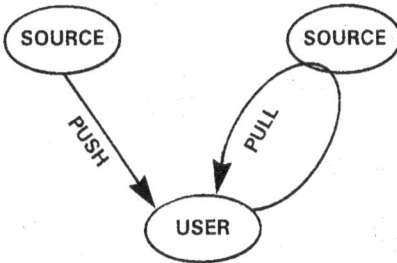

Figure 7. Supply-push and demand-pull information management.

data base, for example—the demand can be filled quickly and efficiently. However, if the information is not readily available, the demand typically triggers a "demand cascade," as the requirement filters through the chain of command until it reaches the appropriate level for gathering. This takes time and can be a burden to lower echelons, especially in a centralized command and control system in which all information must be fed to the senior echelons. An answer to the demand cascade is for commanders to keep dedicated gathering assets which answer directly to them, such as the directed telescopes already mentioned.

Demand-pull can help focus scarce resources on those tasks which the commander has identified as critical; it can deliver information specifically tailored to the commander's information needs; and it will produce only that information which the information needs; and it will produce only that information which the commander requests. These characteristics can be both strengths and weaknesses. They can be strengths because information flow is tailored specifically to identified requirements. However, they can also be weaknesses because there will often be information requirements that the commander has not identified, and in a pure demand-pull system those requirements will go unsatisfied. One definite disadvantage of demand-pull is the cost in time since the search for information may not begin until the commander has identified the need for that information.

We can also discuss information management in terms of how information is transmitted. First, information may be broadcast, sent simultaneously to a broad audience—anyone with access to the information network—to include different echelons of command. (See figure 8.) The great advantage of broadcast is that it gets information to the widest audience in the shortest amount of time. If the information is of a generic nature, this method may be extremely efficient. However, since the information is sent to a wide audience with varying information requirements, the information cannot be tailored to suit any specific commander's needs. Perhaps the greatest drawback of broadcast transmission is that undisciplined use of this method can quickly lead to information overload.

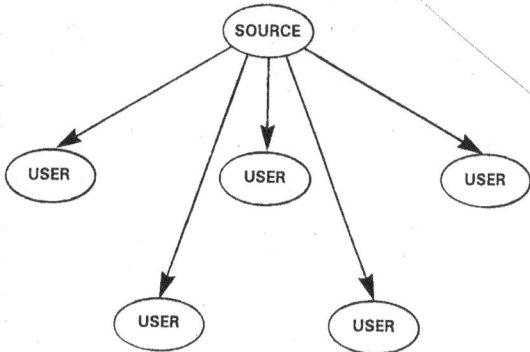

Figure 8. Broadcast transmission.

The alternative to broadcast is point-to-point transmission—or "narrowcast"—in which information is sent to a specific user or users. As appropriate, information is then passed sequentially from one user to the next. (See figure 9.) Point-to-point transmission has two basic advantages. First, information can be tailored to meet the specific needs of each recipient. Second, point-to-point transmission has built-in control mechanisms which broadcast transmission lacks. Each node in the sequence can serve as a control   mechanism, filtering and integrating information as appropriate before passing it on—lessening the risk of overload and tailoring information to the needs of the next recipient. The major disad-

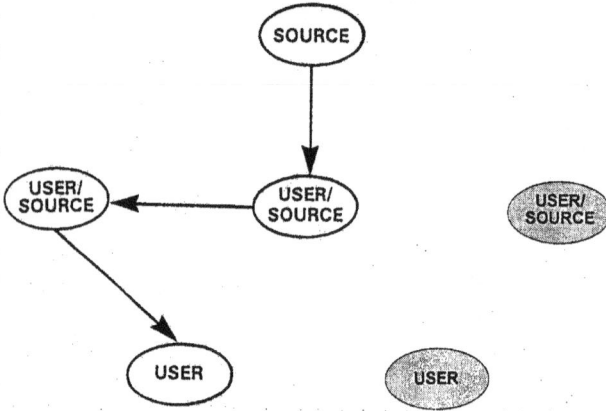

Figure 9. Point-to-point transmission.

vantages of point-to-point are that information reaches a broad audience more slowly and that the chances of distortion increase with each node that information passes through.

In practice, the different aspects of information management are far from incompatible; in fact, combined wisely they can effectively complement one another within the same command and control system.

DECISIONMAKING THEORY

A principal aim of command and control is to enhance the commander's ability to make sound and timely decisions. As we might expect, the defining features of command and control—uncertainty and time—exert a significant influence on decisionmaking.[15] All decisions must be made in the face of uncertainty. Theoretically, we can reduce uncertainty by gaining more information, but any such decrease in uncertainty occurs at the expense of time. And as we have already mentioned, it is not so much the amount of information that matters, but the right elements of information available at the right time and place.

There are two basic theories on how we make decisions.[16] The traditional view is that decisionmaking is an analytical process based on generating several different options, comparing all the options according to some set of criteria, and identifying the best option. The basic idea is that comparing multiple options concurrently will produce the optimal solution. As a result, analytical decisionmaking tends to be methodical and time-consuming. Theoretically, reasoning power matters more than experience.

The other basic approach, called intuitive decisionmaking, rejects the computational approach of the analytical method and instead relies on an experienced commander's (and staff's) intuitive ability to recognize the key elements of a particular problem and arrive at the proper decision. Intuitive decisionmaking thus replaces methodical analysis with an intuitive skill for pattern-recognition based on experience and judgment. The intuitive approach focuses on situation assessment instead of on the comparison of multiple options. Intuitive decisionmaking aims at "satisficing," finding the first solution which will satisfactorily solve the problem, rather than on optimizing, as the analytical approach attempts to do.[17] The intuitive approach is based on the belief that, war being ultimately an art rather than a science, there is no absolutely right answer to any problem. Intuitive decisionmaking works on the further belief that, due to the judgment gained by experience, training, and reflection, the commander will generate a workable first solution, and therefore it is not necessary to generate multiple options. Because it does not involve comparing multiple options, intuitive decisionmaking is generally much faster than analytical decisionmaking. If time permits, the commander may further evaluate this decision; if it proves defective, the commander moves on to the next reasonable solution.

Each approach has different strengths and weaknesses, and determining which approach is better in a given situation depends on the nature of the situation, particularly on how much time and information are available. The analytical approach may be appropriate for prehostility decisions about mobilization or contingency planning when time is not a factor and extensive information can be gathered. It may be useful in situations in which it is necessary to document or justify a decision or in decisions requiring complicated computations which simply cannot be done intuitively (such as in making decisions about supply rates). It may be appropriate when choosing from among several existing alternatives, as in equipment acquisition, for example. Finally, an analytical approach may have some merit in situations in which commanders are inexperienced or in which they face never-be- fore-experienced problems. However, that said, the intuitive approach is more appropriate for the vast majority of typical tactical or operational decisions—decisions made in the fluid, rapidly changing conditions of war when time and uncertainty are critical factors, and creativity is a desirable trait.[18]

We frequently associate intuitive decisionmaking with rapid/time-sensitive planning and analytical decisionmaking with deliberate planning. This may often be the case but not necessarily. For example, a thorough, deliberate planning effort in advance of a crisis can provide the situational awareness that allows a commander to exercise effective intuitive decisionmaking. Conversely, the analytical approach of developing and selecting from several courses of action may be done rapidly. The point is that the planning model or process we choose, and the decisionmaking approach that supports it, should be based upon the situation, the time available, the knowledge and situational awareness of the organization, and the commander's involvement in the planning and decisionmaking process. While the two approaches to decisionmaking are conceptually distinct, they are rarely mutually exclusive in practice.

CONCLUSION

Our view of the true nature of war leads us to one of two responses to dealing with the fundamental problem of command: either pursuing certainty or coping with uncertainty. These responses lead to two distinctly different theories of command and control. Each theory in turn imposes its own requirements on the various aspects of command and control—decisionmaking, communications, information management, planning, organization, training, education, doctrine, and so on—and so forms the basis for a distinct and comprehensive approach to command and control. The question is: Which approach do we adopt? The Marine Corps' concept of command and control is based on accepting uncertainty as an undeniable fact and being able to operate effectively despite it. The Marine Corps' command and control system is thus built around mission command and control which allows us to create tempo, flexibility, and the ability to exploit opportunities but which also requires us to decentralize and rely on low-level initiative. In the next chapter, we will discuss the features of such a command and control system.

Chapter 3

Creating Effective Command and Control

"Whoever can make and implement his decisions consistently faster gains a tremendous, often decisive advantage. Decision making thus becomes a time-competitive process, and timeliness of decisions becomes essential to generating tempo."

—FMFM 1, *Warfighting*

Having reached a common understanding of the nature of command and control and having laid out its key theories, we can develop the characteristics of an effective command and control system. How do we create effective command and control, both in our units and within the Marine Corps as a whole?

THE CHALLENGES TO THE SYSTEM

Before we discuss the features of our command and control system, it might help to review the challenges that the system, as a complex blend of people, information, and support, must face. What obstacles must our command and control system overcome and what must it accomplish? First and foremost, the system must deal effectively with the twin problems of uncertainty and time. It must be compatible with our doctrine of maneuver warfare. It must function effectively across a broad spectrum of conflicts and environments—that is, in "any clime and place." Moreover, while designed principally to work effectively in war, it should also apply to peacetime activities, operational or administrative.

Our command and control must improve our ability to generate a higher tempo of action than the enemy. It should help us adapt to rapidly changing situations and exploit fleeting opportunities. It should allow us to withstand disruptions of all kinds, created by the enemy, the environment, or our- selves, since we recognize that disruption will be a normal course of events. It should help to gather information quickly, accurately, and selectively and to get the right information to the right person at the right time and in the right form—with- out creating information overload. It should improve our abil- ity to build and share situational awareness.

Our command and control should help provide insight into the nature of the problem facing us and into the nature and de- signs of our enemy. It should help us to identify critical en- emy vulnerabilities and should provide the means for focus- ing our efforts against those vulnerabilities. At the same time, it should help conceal our true designs from the enemy. It should help establish goals which are both meaningful and practicable, and it should help devise workable, flexible plans to accom- plish those goals.

It should facilitate making timely and sound decisions de- spite incomplete and unclear information, and it should provide the means to modify those decisions quickly. It should allow us to monitor events closely enough to ensure proper execution, yet without interfering with subordinates' actions. It should help us communicate instructions quickly, clearly, and con- cisely and in a way that provides subordinates the necessary guidance without inhibiting their initiative.

With this in mind, what should such a command and control system look like?

108

MISSION COMMAND AND CONTROL

First and foremost, our approach should be based on mission command and control. Mission command and control is central to maneuver warfare. We realize that the specific combination of command and control methods we employ in a particular situation depends on the unique requirements of that situation. We also realize that, within an overall mission approach, detailed command and control may be preferable for certain procedural or technical tasks. That said, however, for the overall command and control of military actions, we should use mission command and control as much as the situation allows. Why? Mission command and control deals better with the fundamental problems of uncertainty and time. Since we recognize that precision and certainty are unattainable in war anyway, we sacrifice them for speed and agility. Mission command and control offers the flexibility to deal with rapidly changing situations and to exploit fleeting windows of opportunity. It provides for the degree of cooperation necessary to achieve harmony of effort yet gives commanders at all levels the latitude to act with initiative and boldness.

Mission command and control relies on the use of *mission tactics* in which seniors assign missions and explain the underlying intent but leave subordinates as free as possible to choose the manner of accomplishment. Commanders seek to exercise a sort of *command by influence*, issuing broad guidance rather than detailed directions or directives. The higher the level of command, the more general should be the supervision and the less the burden of detail. Commanders reserve the use of close personal supervision to intervene in subordinate's actions only in exceptional cases. Thus all commanders in their own spheres are accustomed to the full exercise of authority and the free application of judgment and imagina- tion.[1] Mission command and control thus seeks to maximize low-level initiative while achieving a high level of cooperation in order to obtain better battlefield results.

Orders should include restrictive control measures and should prescribe the manner of execution only to the degree needed to provide necessary coordination that cannot be achieved any other way. Orders should be as brief and as simple as possible, relying on subordinates to work out the details of execution and to effect the necessary coordination. Mission command and control thus relies on lateral coordination between units as well as communications up and down the chain.

The aim is not to increase our *capacity* to perform command and control. It is not more command and control that we are after. Instead, we seek to decrease the amount of com- mand and control that we *need.* We do this by replacing coercive command and control methods with spontaneous, self-disciplined cooperation based on low-level initiative, a commonly understood commander's intent, mutual trust, and implicit understanding and communications.

LOW-LEVEL INITIATIVE

Initiative is an essential element of mission command and control since subordinates must be able to act without instructions. Our warfare doctrine emphasizes seeking and rapidly exploiting fleeting opportunities, possible only through low-level initiative. Initiative hinges on distributing the authority to decide and act throughout an organization rather than localizing it in one spot. And as we have already discussed, where there is authority, there is also responsibility. Being free to act on their own authority, subordinates must accept the corresponding responsibility to act.

Our command and control must be biased toward decision and action at all levels. Put another way, the command and control process must be self-starting at every level of command as all commanders within their own spheres act upon the need for action rather than only on orders from above.

It is important to point out that initiative does not mean that subordinates are free to act without regard to guidance from above. In fact, initiative places a special burden on subordinates, requiring that they always keep the larger situation in mind and act in consonance with their senior's intent. The freedom to act with initiative thus implies a greater obligation to act in a disciplined and responsible way. Initiative places a greater burden on the senior as well. Delegating authority to subordinates does not absolve higher command-ers of ultimate responsibility. They must frame their guidance in such a way that provides subordinates sufficient understanding to act in consonance with their desires while not restricting freedom of action. Commanders must be adept at expressing their desires clearly and forcefully—a skill that requires practice.

Beyond its tactical utility, initiative has an important psychological effect on the members of an organization. Recognizing what needs to be done and taking the action necessary to succeed is a satisfying experience and a powerful stimulant to human endeavor. People not merely carrying out orders but acting on their own initiative feel a greater responsibility for the outcome and will naturally act with greater vigor. Thus, initiative distributed throughout is a source of great strength and energy for any organization, especially in times of crisis.[2]

As we emphasize initiative, we must recognize that subordinates will sometimes take unexpected actions, thus imposing on commanders a willingness to accept greater uncertain- ty with regard to the actions of their subordinates.

COMMANDER'S INTENT

In a decentralized command and control system, without a common vision there can be no unity of effort; the various actions will lack cohesion. Without a commander's intent to express that common vision, there simply can be no mission command and control.

There are two parts to any mission: the task to be accomplished and the reason, or intent. The task describes the action to be taken while the intent describes the desired result of the action. Of the two, the intent is predominant. While a situation may change, making the task obsolete, the intent is more enduring and continues to guide our actions. Understanding our commander's intent allows us to exercise initiative in harmony with the commander's desires.

The commander's intent should thus pull the various separate actions of the force together, establishing an underlying purpose and focus. It should provide topsight. In so doing, it should provide the logic that allows subordinates each to act according to their unique circumstances while maintaining harmony with one another and the higher commander's aim. While assigned tasks may be overcome by events, the commander's intent should allow subordinates to act with initiative even in the face of disorder and change.

In a system based on mission command and control, providing intent is a prime responsibility of command and an essential means of leading the organization.

MUTUAL TRUST

Mission command and control demands mutual trust among all commanders, staffs, and Marines—confidence in the abilities and judgment of subordinates, peers, and seniors. Trust is the cornerstone of cooperation. It is a function of familiarity and respect. A senior trusts subordinates to carry out the assigned missions competently with minimal supervision, act in consonance with the overall intent, report developments as necessary, and effect the necessary coordination. Subordinates meanwhile trust that the senior will provide the necessary guidance and will support them loyally and fully, even when they make mistakes.

Trust has a reverse side: it must be earned as well as given. We earn the trust of others by demonstrating competence, a sense of responsibility, loyalty, and self-discipline. This last is essential. Discipline is of fundamental importance in any military endeavor, and strict military discipline remains a pillar of command authority. But since mission command and control is decentralized rather than centralized and spontaneous rather than coercive, discipline is not only imposed from above; it must also be generated from within. In order to earn a senior's trust, subordinates must demonstrate the self-discipline to accomplish the mission with minimal supervision and to act always in accord with the larger intent. Seniors, in order to earn subordinates' trust, must likewise demonstrate that they will provide the subordinate the framework within which to act and will support and protect subordinates in every way as they exercise initiative.

Mutual trust also has a positive effect on morale: it increases the individual's identity with the group and its goals. Mutual trust thus contributes to a supportive, cooperative environment.

IMPLICIT UNDERSTANDING AND COMMUNICATION

The final essential ingredients of effective mission command and control are implicit understanding and communication which are the basis for cooperation and coordination in maneuver warfare.[3] These intangible human abilities allow us to harmonize our actions intuitively with others.

Implicit understanding and communication do not occur automatically. They are abilities we must actively foster and are the product of a common ethos and repeated practice—as with the members of a basketball team who think and move as one or the members of a jazz band who can improvise freely without losing their cohesion. Gaining this special state of organizational effectiveness has significant implications for doctrine, education, and training, as we will discuss.

DECISIONMAKING

Effective decisionmaking at all levels is essential to effective command and control. Several general principles apply. First, since war is a clash between opposing wills, all decisionmaking must first take our enemies into account, recognizing that while we are trying to impose our will on them, they are trying to do the same to us. Second, whoever can make and implement decisions faster, even to a small degree, gains a tremendous advantage. The ability to make decisions quickly on the basis of incomplete information is essential. Third, a military decision is not merely the product of a mathematical computation, but requires the intuitive and analytical ability to recognize the essence of a given problem and the creative ability to devise a practical solution. All Marine decisionmakers must demonstrate these intuitive, analytical, and creative skills which are the products of experience, intelligence, boldness, and perception. Fourth, since all decisions must be made in the face of uncertainty and since every situation is unique, there is no perfect solution to any battlefield problem; we should not agonize over one. We should adopt a promising scheme with an acceptable degree of risk, and do it more quickly than our foe. As General George Patton said, "A good plan violently executed *Now* is better than a perfect plan next week."[4] And finally, in general, the lower the eche-

lon of command, the faster and more direct is the decision process. A small-unit leader's decisions are based on factors usually observed firsthand. At successively higher echelons of command, the commander is further removed from events by time and distance. As a consequence, the lower we can push the decisionmaking threshold, the faster will be the decision cycle.

Maneuver warfare requires a decisionmaking approach that is appropriate to each situation. We must be able to adopt and combine the various aspects of both intuitive and analytical decisionmaking as required. Because uncertainty and time will drive most military decisions, we should emphasize intuitive decisionmaking as the norm and should develop our leaders accordingly. Emphasizing experienced judgment and intuition over deliberate analysis, the intuitive approach helps to generate tempo and to provide the flexibility to deal with uncertainty. Moreover, the intuitive approach is consistent with our view that there is no perfect solution to battlefield problems and with our belief that Marines at all levels are capable of sound judgment. However, understanding the factors that favor analytical decisionmaking—especially when time is not a critical factor—we should be able to adopt an analytical approach or to reinforce intuitive decisionmaking with more methodical analysis.

INFORMATION MANAGEMENT

Our management of information should facilitate the rapid, distributed, and unconstrained flow of information in all directions. At the same time, it should allow us to discriminate as to importance, quality, and timeliness as a means of providing focus and preventing information overload. It should enhance the ability of all commanders to communicate a concept and intent with clarity, intensity, and speed.

We should supply information, as much as possible, in the form of meaningful images rather than as masses of data. This means, among other things, that our system must have the means of filtering, fusing, and prioritizing information. By filtering we mean assessing the value of information and culling out that which is not pertinent or important. By fusing we mean integrating information into an easily usable form and to an appropriate level of detail. And by prioritizing we mean expediting the flow of information according to importance. All information management should focus on critical information requirements. This demands vision on the part of the commander and understanding on the part of subordinates in order to recognize critical information when they see it.

Our command and control system should make use of all the various channels and methods by which information flows—implicit as well as explicit and informal as well as formal. Our system must facilitate communications in all directions, not only vertically within the chain of command, and should ensure that information flow is interactive rather than one-way. Our system should provide redundant channels as a safeguard against disruption and battle damage; which channel information follows is less important than whether it reaches the right destination.

Since information is changed by each person who handles it, important information should pass directly between principal users, eliminating intermediaries, such as equipment operators or clerks. Wherever possible, person-to-person information should be communicated by word of mouth and face-to-face since humans communicate not only by what they say but also by how they say it. The desire to have principals communicate directly and by voice does not mean that we do not need to keep a record of communications; permanent records can be important as a means of affirming understanding and for reasons of later study and critique.

Our information management system should be a hybrid exhibiting the judicious combination of broadcast and point-to-point transmission and supply-push and demand-pull.[5]

Generic information of value to many users at a variety of echelons may be broadcast, the transmission method which reaches the broadest audience most quickly. However, we must exercise discipline with respect to broadcast transmissions to avoid the danger of information overload. In comparison, we should use point-to-point transmission for information that needs to be tailored to suit the needs of individual users.

Our information management system should also combine the best characteristics of supply-push and demand-pull. We recognize that supply-push is the most efficient way to provide much of the information needed routinely—whether broadcast or point-to-point. Through the implicit understanding and shared images of its members, the system should attempt to anticipate commanders' needs and should attempt to push routine information to an easily accessible, local data base. Commanders then pull from the base only that information they need. In this way, we avoid the danger of information overload associated with supply-push and broadcast and circumvent some of the delays normally associated with demand-pull. (See figure 10.)

We also recognize that commanders will likely be unaware of the need for certain information, so we must ensure that truly critical, time-sensitive information is pushed directly to them without delay, even if it means skipping intermediate echelons of command. Echelon-skipping does not mean, however, that intermediate echelons are left uninformed. After critical information has passed directly between the primarily concerned echelons, both those echelons should inform intermediates by normal channels.

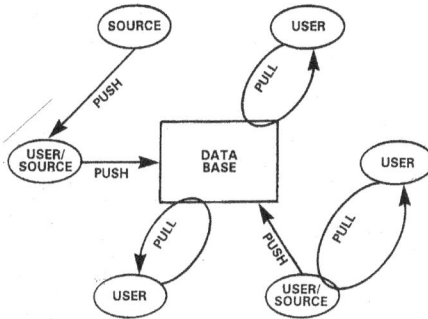

Figure 10. Hybrid information management system.

Additionally, since no system can effectively anticipate all information needs, commanders must have available directed telescopes by which they can satisfy their own information needs quickly. It is important, however, that the directed telescope not interfere (or be perceived to interfere) with the normal functioning of the chain of command: the perception of spying or intruding on the province of subordinate commanders can damage the vital trust between senior and subordinate.

LEADERSHIP

Because people are the first and most important element of our command and control system, strong and effective leadership is of essential importance to our command and control. Mission command and control requires predominantly a persuasive or delegating approach to leadership. It becomes the role of the leaders to motivate Marines to perform to the highest standards and to instill self-discipline. Leadership is thus a matter of enlisting the committed, enthusiastic, and loyal performance of subordinates, both as a habitual quality and in support of each specific mission. A good part of this comes from keeping subordinates informed and from one's own demonstration of commitment.

Mission command and control does not imply a depersonalized or detached demeanor. Since leaders who employ mission tactics are not overly involved in the details of execution, one of their primary roles will thus be to provide the intent which holds together the decentralized actions of subordi- nates.

Another important role of leadership will be to create a close-knit sense of team which is essential to developing trust and understanding within the organization. Leaders should reinforce the common core values which are the basis for implicit understanding and trust. Leaders should strive to create an atmosphere of mutual support in which subordinates are encouraged to demonstrate initiative and to effect the necessary coordination locally. Importantly, this means avoiding a "zero-defects" mentality which tends to penalize initiative.

Leadership also becomes a matter of developing subordinate maturity—which means engendering in subordinates a willingness to exercise initiative, the judgment to act wisely, and an eagerness to accept responsibility. Leader-as-teacher is an essential component of our approach to leadership.

PLANNING

Planning is an essential element of effective command and control. Our philosophy of command and control calls for planning methods that are based on the particulars of each situation, especially on the nature of the activity being planned. In general, we should not think of planning as a scripting process which establishes specific actions to be taken and often establishes timetables for those actions. This approach seeks to narrow possibilities in order to minimize uncertainty and simplify preparations and coordination. Rather, we should view planning as a learning process which helps us understand how to exploit the various possibilities an uncertain future may hold. The intent should be to maximize opportunities in order to generate freedom of action and not to minimize possibilities in order to simplify coordination.

Whether done rapidly or deliberately, effective planning requires a sensitive awareness and judicious use of time. If time is available, there can be little excuse for not planning adequately. A company commander who spends an hour deliberately developing a detailed plan in the heat of a crisis when seconds matter is no better than a division commander who has several days to prepare for an amphibious landing and hastily develops an ill-conceived ship-to-shore plan. Just because time may be available does not mean that we should use it to develop lengthy, detailed directives. Elaborateness and detail are not generally measures of effective plans. Instead, directives should convey the *minimum* amount of instruction necessary for execution. *Directives should be as clear, simple, and concise as each situation permits.*

Planning should be participatory. The main benefits of planning are not from consuming the product but from engaging in the process. In other words, the planning matters more than the plan. We should view any plan as merely a common starting point from which to adapt as the situation requires and not as a script which must be followed. We should think of the plan as a scheme for solving a problem. Since the future will always be uncertain, plans must be flexible and adaptable, allowing the opportunity to pursue a variety of options.

Effective planning must involve an appreciation for time horizons. We must project far enough into the future so that we can maintain the initiative and prepare adequately for upcoming action, but not so far into the future that plans will have little in common with actual developments. Effective planning should facilitate shaping the conditions of the situation to our advantage while preserving freedom to adapt quickly to real events. As actions approach and our ability to influence them grows, planning should have helped us develop an appreciation for the situation and get into a position to exploit it.

As with decisionmaking, we should decentralize execution planning to the lowest possible levels *so that those who must execute have the freedom to develop their own plans.* A plan should dictate a subordinate's actions only to the minimum degree essential to provide necessary coordination unattainable any other way. Ideally, rather than dictating a subordinate's actions, a good plan should actually create opportuni- ties for the subordinate to act with initiative.

Without question, planning is an important and valuable part of command and control. However, we must guard against overcontrol and mechanical thinking. A properly framed commander's intent and effective commander's planning guidance create plans which foster the environment for subordinate commanders to exercise initiative to create tempo while allowing for flexibility within execution of operations. The object of planning is to provide options for the commander to face the future with confidence. The measure of a good plan is not whether it transpires as designed but whether it facilitates effective action in the face of unforeseen events.

FOCUSING COMMAND AND CONTROL

The focus of the command and control effort should reflect the overall focus of efforts. We should focus the command and control effort on critical tasks and at critical times and places. We can do this by a variety of means. We concentrate information-gathering assets and other command and control resources where they are needed most. We concentrate planning, coordination, analysis, and other command and control activities on the most important tasks, and we exercise economy elsewhere. We prioritize information requirements and concentrate gathering, processing, and communications on the critical elements. We filter, prioritize, and fuse information to ensure that critical, time-sensitive information moves quickly and effectively and that less important information does not clutter communications channels. We manage that most precious of all commodities, time, to ensure that the most important tasks receive our earliest and utmost attention. We especially ensure that commanders devote their time and energies only to critical tasks, and that they are protected against routine distractions. The commander should do only those things which only the commander can do or which nobody else can do adequately. Routine tasks must be delegated to others.

A key way commanders can provide focus is by personal attention and presence. In the words of Field Marshal Sir William Slim, "One of the most valuable qualities of a commander is the flair for putting himself in the right place at the vital time." As we have mentioned, by positioning themselves at the critical spot, commanders can observe events more directly and avoid the delays and distortions that occur as information filters up the chain of command. In so doing, commanders can gain firsthand the essential appreciation for the situation which can rarely be gained any other way. Equally important, they can *influence* events more directly and avoid the delays and distortions that occur as information filters *down* the chain of command. By their personal presence, commanders can provide the leadership that is so essential to success in war. And simply by the moral authority that their presence commands, commanders direct emphasis to the critical spot and focus efforts on the critical task.

We have discussed the need to gain several different images. Commanders go wherever they must to get the most important image. For the closeup image, this often means at the front—which does not necessarily mean at the forwardmost point of contact on the ground, but wherever the critical action is taking place or the critical situation is developing. For ground commanders, even senior ones, this may in fact mean at or near the point of contact. But for others, and even for ground commanders, this may mean with a subordinate commander in the critical sector—in a ready room listening to flight debriefs during an important phase of an air operation, at a critical point along a route of march, or in an aircraft flying over the battlefield. If the critical view at a particular moment is the overall picture, the commander may want to be in the command post's operations center, piecing together various reports from far-flung sources, or even at a higher headquarters, learning about the larger situation (although in general it is better for senior commanders to come forward than for subordinate commanders to go rearward to exchange information). And for that matter, if a commander is trying to get inside the mind of an adversary who has made a bold and unexpected move that has shattered situational awareness, the best place may be sequestered from distractions, sitting against a tree, alone with a map.

Our philosophy of command calls for energetic and active commanders with a flair, as Slim says, for being in the critical place, lending leadership, judgment, and authority wherever it is needed most. The commander might start at the command post to piece together an overall image and supervise the development of the plan, but should then usually move forward to supervise execution at the critical spot, returning to the command post only long enough to regenerate an image of the overall situation before moving out again to the next critical spot. The important point is that commanders must not feel tied to the command post, unable to leave it for fear of missing a valuable report—especially since modern communications increasingly allow commanders to stay informed even when away from the command post. When commanders leave the command post, it is imperative that they empower the staff to act on their behalf. The staff must be able to act with initiative when the commander is away and therefore must understand the commander's estimate of the situation, overall intent, and designs. Mutual trust and implicit understanding apply to the staff as much as to subordinate commanders. Commanders who do not empower the staff to act on their behalf will become prisoners in their own headquarters, out of touch with reality and limited in their ability to influence events.

THE COMMAND AND CONTROL SUPPORT STRUCTURE

It is important to keep in mind that the command and control support structure merely provides the supporting framework for our command and control; it does not constitute the system itself. The sole purpose of the support structure is to assist people in recognizing what needs to be done and in taking appropriate action. In addition to supporting our approach to command and control, the components of our command and control support structure must be compatible with one another. And since people are the driving element behind command and control, the components of the structure, together and alone, must be user-friendly—that is, designed first and always with people in mind.

TRAINING, EDUCATION, AND DOCTRINE

Collectively, training, education, and doctrine prepare people for the roles they play in command and control. First, since mission command and control demands initiative and sound decisionmaking at all levels, training, education, and doctrine must aim at fostering initiative and improving decisionmaking ability among all Marines. It is not enough to allow initiative; we must actively encourage and demand an eagerness to accept responsibility. This means that we must develop an institutional prejudice for tolerating mistakes of action but not inaction. Training and education should seek to develop in leaders the pattern-recognition skills that are essential to intuitive decisionmaking.

Second, training, education, and doctrine must prepare Marines to function effectively in varying environments amid uncertainty and disorder and with limited time. Exercise scenarios should purposely include elements of disorder and uncertainty—an unexpected development or mission change, as examples. Field exercises and command post exercises should purposely include disruption of command and control, for example, "destruction" of a main command post or loss of communications during a critical phase of an evolution. Planning exercises should incorporate severe time limits to simulate stress and tempo. As Field Marshal Erwin Rommel said, "A commander must accustom his staff to a high tempo from the outset, and continuously keep them up to it."

Third, education and training should teach the appropriate use of techniques and procedures. Training should provide techniques and procedures which emphasize flexibility, speed, and adaptability—fast and simple staff planning models, for example. Education should provide an understanding of when to apply different techniques and procedures—when to use intuitive or analytical decisionmaking techniques, for example.

Last and perhaps most important, training, education, and doctrine should provide a shared ethos, common experiences, and a shared way of thinking as the basis for the trust, cohesion, and implicit communication that are essential to maneuver warfare command and control. They should establish a common perspective on how Marines approach the problems of command and control.

PROCEDURES

Used properly, procedures can be a source of organizational competence—by improving a staff's efficiency or by increasing planning tempo, for example. Procedures can be especially useful to improve the coordination among several people who must cooperate in the accomplishment of repetitive tasks—such as the internal functioning of a combat operations center. Used improperly, however, procedures can have the opposite effect: applied blindly to the wrong types of tasks or the wrong situations, they can lead to ineffective, even dysfunctional performance.

We must recognize that procedures apply only to rote or mechanical tasks. They are not acts of judgment, nor are they meant to replace the need for judgment. The purpose of procedures "is not to restrict human judgment, but to free it for the tasks only it can perform." [6] We must keep in mind that procedures are merely tools to be used, modified, or discarded as the situation requires. They are not rules which we must follow slavishly.

Our command and control procedures should be designed for simplicity and speed. They should be designed for simplicity so that we can master them easily and perform them quickly and smoothly under conditions of extreme stress. They should be designed for speed so that we can generate tempo. Streamlined staff planning sequences, for example, are preferable to deliberate, elaborate ones. The standard should be simple models which we can expand if time and circumstances permit, rather than inherently complicated models which we try to compress when time is short—which is likely to be most of the time. As Second World War German General Hermann Balck used to say to his staff, "Don't work hard, work fast."

MANPOWER

Since people are the first and driving element of our command and control system, effective manpower management is essential to command and control. Since mission command and control relies heavily on individual skills and judgment, our manpower management system should recognize that all Marines of a given grade are not interchangeable and should seek to put the right person in the right billet based on specific ability and temperament. Additionally, the manpower management system should seek to achieve personnel stability within units and staffs as a means of fostering the cohesion, teamwork, and implicit understanding that are vital to mission command and control. We recognize that casualties in war will take a toll on personnel stability, but the greater stability a unit has initially, the better it will absorb those casualties and incorporate replacements.

ORGANIZATION

The general aims of organization with regard to command and control should be to create unity of effort, reasonable spans of control, cohesive mission teams, and effective information distribution. Organization should not inhibit communications in any way but instead should facilitate the rapid distribution of information in all directions and should provide feedback channels.

In general, we should take a flexible approach to organization, maintaining the capability to task-organize our forces to suit the situation which might include the creation of nonstandard and temporary task groupings. However, the commander must reconcile this desire for organizational flexibility with the need to create implicit understanding and mutual trust which are the product of familiarity and stable working relationships.

Mission command and control requires the creation of self-reliant task groups capable of acting semiautonomously. By task-organizing into self-reliant task groups, we increase each commander's freedom of action and at the same time decrease the need for centralized coordination of support.

We should seek to strike a balance between "width" and "depth" so that the organization is suited to the particular situation. The aim is to flatten the organization to the greatest extent compatible with reasonable spans of control. Commanders should have the flexibility to eliminate or bypass selected echelons of command or staff as appropriate in order to improve operational tempo. Additionally, it is not necessary that all echelons of command exercise all functions of command. Just as we task-organize our force, so should we task-organize our command and control structure.

A word is in order about the size of staffs. The larger and more compartmented the staff, the more information it requires to function. This increase in information in turn requires an even larger staff, and the result is a spiraling increase in size. However, the larger a command and control organization, the longer it generally takes that organization to perform its functions. In the words of General William T. Sherman, "A bulky staff implies a division of responsibility, slowness of action and indecision, whereas a small staff implies activity and concentration of purpose." [7] Also, a large staff takes up more space, emits a larger electromagnetic signature, and is less mobile than a small one, and consequently is more vulnerable to detection and attack. A large staff, with numerous specialists, may be more capable of detailed analysis and planning than a small one, but we have already established that we generally value speed and agility over precision and certainty. We should therefore seek to keep the size of staffs to a minimum in order to facilitate a high operating tempo and to minimize the space and facilities that the headquarters requires. The ideal staff would be so austere it could not exercise fully detailed command and control.

EQUIPMENT AND TECHNOLOGY

Equipment, to include facilities, is an integral part of any command and control support structure, but we must re- member that it is only one component. As we have mentioned, there are two dangers in regard to command and control equipment, the first being an overreliance on technology and the second being a failure to make proper use of technological capabilities. The aim is to strike a balance that gets the most out of our equipment and at the same time integrates technology properly with the other components of the system.

We believe very strongly that the object of technology is not to reduce the role of people in the command and control process, but rather to enhance their performance—although technology should allow us to decrease the *number* of people involved in the process. As a first priority, equipment and facilities should be user-friendly. Technology should seek to automate routine functions which machines can accomplish more efficiently than people in order to free people to focus on the aspects of command and control which require judgment and intuition. We may even use technology to assist us in those human activities so long as we do not make the mistake of trying to replace the person who can think with the machine that cannot.

Command and control equipment should help improve the flow and value of information within the system. But as we have said repeatedly, improving information is not simply a matter of increasing volume; it is also a matter of quality, timing, location, and form. To the greatest extent possible, communications equipment should connect principals directly, minimizing the need for specialized operators. Ad- ditionally, equipment should minimize the input burden placed on people; ideally, the input of information into the system should be automatic. Last, but hardly least, technological developments should focus on presenting information in a way that is most useful to humans—that is, in the form of meaningful visual images rather than lists of data.

As with all the components of our command and control support structure, our command and control equipment should be consistent with our overall approach to command and control. For example, equipment that facilitates or encourages the micromanagement of subordinate units is inconsistent with our command and control philosophy. Moreover, such technological capability tends to fix the senior's attention at too low a level of detail. A regimental commander, for example, does not as a rule need to keep track of the movements of every squad (although with position-locating technology it may be a temptation); a regimental commander needs a more general appreciation for the flow of action. Commanders who focus at too low a level of detail (whether because the technology tempts them to or not) risk losing sight of the larger picture.

The reality of technological development is that equipment which improves the ability to monitor what is happening may also increase the temptation and the means to try to direct what is happening. Consequently, increased capability on the part of equipment brings with it the need for increased understanding and discipline on the part of users. Just because our technology allows us to micromanage does not mean that we should.

CONCLUSION

Our approach to command and control recognizes and accepts war as a complex, uncertain, disorderly, and time-competitive clash of wills and seeks to provide the commander the best means to win in that environment. We seek to exploit trust, co-operation, judgment, focus, and implicit understanding to lessen the effects of the uncertainty and friction that are conse-quences of war's nature. We rely on mission command and control to provide the flexibility and responsiveness to deal with uncertainty and to generate the tempo which we recognize is a key element of success in war. We focus on the value and timeliness of information, rather than on the amount, and on getting that information to the right people in the right form. We seek to strike a workable balance among people, proce-dures, and technology, but we recognize that our greatest com-mand and control resource is the common ethos and the resulting bond shared by all Marines.

Notes

Epigraphs: The quotation on page **33** is from Carl von Clausewitz, *On War*, Book 1, chapter 6. The quotation on page **61** is from Martin van Creveld, *Command in War,* p. 269. The quotation on page **105** is from FMFM 1, *Warfighting*, p. 69.

The Nature of Command and Control

1. Command and control the business of the commander: In Joint Pub 1-02, *Department of Defense Dictionary of Military and Associated Terms*, defined as: "The exercise of authority and direction by a properly designated commander over assigned forces in the accomplishment of the mission. Command and control functions are performed through an arrangement of personnel, equipment, communications, facilities, and procedures employed by a commander in planning, directing, coordinating, and controlling forces and operations in the accomplishment of the mission."

2. Authority and responsibility: Henri Fayol, *General and Industrial Management* (Pitman Publishing Corp., 1949), pp. 21–22.

3. The traditional view: Command in Joint Pub 1-02 is defined as "the authority that a commander in the Military Service lawfully exercises over subordinates by virtue of rank or assignment. Command includes the authority and responsibility for effectively using available resources and for planning the employment

of, organizing, directing, coordinating, and controlling military forces for the accomplishment of assigned missions. It also includes responsibility for health, welfare, morale, and discipline of assigned personnel." Control in Joint Pub 1-02 is defined in our context as "physical or psychological pressures exerted with the intent to assure that an agent or group will respond as directed."

4. Feedback as control: See Norbert Wiener, *Cybernetics, or, Control and Communication in the Animal and the Machine*, 2d ed. (Cambridge, MA: MIT Press, 1962), pp. 95–115, and *The Human Use of Human Beings: Cybernetics and Society* (Boston: Houghton Mifflin, 1950), pp. 12–15 and pp. 69–71. As applied to command and control: See John R. Boyd, "An Organic Design for Command and Control," *A Discourse on Winning and Losing*, unpublished lecture notes, 1987.

5. The illusion of being "in control": Peter M. Senge, *The Fifth Discipline: The Art and Practice of The Learning Organization* (New York: Doubleday/Currency, 1990), pp. 190–193.

6. Command and control as a complex (adaptive) system: See M. Mitchell Waldrop, *Complexity: The Emerging Science at the Edge of Order and Chaos* (New York: Simon & Schuster, 1992); Roger Lewin, *Complexity: Life on the Edge of Chaos* (New York: Macmillan, 1992); or Kevin Kelly, *Out of Control: The New Biology of Machines: The Rise of Neo-Biological Civilization* (Reading, MA: Addison-Wesley, 1994). Also described as "far-from-equilibrium, nonlinear" systems in Ilya Prigogine and Isabelle Stengers, *Order Out of Chaos: Man's New Dialogue with Nature* (New York: Bantam Books, 1984) and Gregoire Nicolis and Ilya Prigogine, *Exploring Complexity: An Introduction* (New York: W.H. Freeman & Co., 1989).

7. "Success is not due simply to general causes . . . ": Carl von Clausewitz, *On War*, trans by Michael Howard and Peter Paret (Princeton, NJ: Princeton University Press, 1984), 595.

8. "Organic" versus "mechanistic" systems: T. Burns, "Mechanistic and Organismic Structures," in Derek Salman Pugh, comp., *Organization Theory: Selected Readings* (Harmondsworth, England: Penguin Books, 1971), pp. 43–55; David K. Banner and T. Elaine Gagné, *Designing Effective Organizations: Traditional & Transformational Views* (Thousand Oaks, CA: Sage Publications, 1995), pp. 152–194; Gareth Morgan, *Images of Organization* (Beverly Hills, CA: Sage Publications, 1986).

9. "Command" and "control" as nouns and verbs: Thomas P. Coakley, *Command and Control for War and Peace* (Washington: National Defense University Press, 1992), p.17.

10. Information as a control parameter: Jeffrey R. Cooper, "Reduced Instruction Set Combat: Processes & Modeling." Presentation given at Headquarters Marine Corps, 5 Jan 95.

11. Command and control support structure: In Joint Pub 1-02: "Command and control system—The facilities, equipment, communications, procedures, and personnel essential to a commander for planning, directing, and controlling operations of assigned forces pursuant to the missions assigned."

12. Coakley, p. 17.

13. Uncertainty as the defining feature of command: See Martin van Creveld, *Command in War* (Cambridge, MA: Harvard University Press, 1985), especially chapters 1 and 8.

14. "War is the realm of uncertainty . . .": Carl von Clausewitz, *On War*, p. 101.

15. Uncertainty as doubt which blocks action: See Ra'anan Lipshitz and Orna Strauss, "Coping with Uncertainty: A Naturalistic Decision Making Analysis," unpublished paper, 1996.

Command and Control Theory

1. The OODA loop: John R. Boyd, "Patterns of Conflict" and "An Organic Design for Command and Control," *A Discourse on Winning and Losing*. The OODA loop is, naturally, a simplification of the command and control process (since we have already described command and control as a process characterized by feedback and other complex interactions). It is not meant to provide a complete description of the various phases and interactions, but rather a basic conceptual model. Numerous individual interactions take place within and among each of the four basic steps. Any effort to divide a complex process like command and control into neat, sequential steps is necessarily going to be partly artificial. Various other similar command and control models exist. We have selected the Boyd model because it is widely known to many Marines. See also William S. Lind, *Maneuver Warfare Handbook* (Boulder, CO: Westview Press, 1985), pp. 4–6.

2. The information (cognitive) hierarchy: Jeffrey R. Cooper, "The Coherent Battlefield—Removing the 'Fog of War.' " Unpublished paper, SRS Technologies, June 1993. Also Cooper, "Reduced Instruction Set Combat: Processes and Modeling."

3. Not only do people think in images, they understand best and are inspired most . . . : Thomas J. Peters, *Thriving on Chaos: Handbook for a Management Revolution* (New York: Alfred A. Knopf, 1988), p. 418.

4. Gavish: "There is no alternative . . .": quoted in Martin van Creveld, *Command in War*, p. 199.

5. "Topsight": David Hillel Gelernter, *Mirror Worlds, or, The Day Software Puts the Universe in a Shoebox–How It Will Happen and What It Will Mean* (New York: Oxford University Press, 1991), pp. 51–53. Gelernter argues that topsight is "the most precious intellectual commodity known to man. . . . It is *the* quality that distinguishes genius in any field." (Italics in original.)

6. The directed telescope: Van Creveld, *Command in War*, p. 75 and pp. 255–57. See also Gary B. Griffin, *The Directed Telescope: A Traditional Element of Effective Command*, Combat Institute Studies Report No. 9 (Ft. Leavenworth, KS: Combat Studies Institute, U.S. Army Command and General Staff College, 1985).

7. Control as "coercive" or "spontaneous": Gregory D. Foster, "Contemporary C^2 Theory and Research: the Failed Quest for a Philosophy of Command," *Defense Analysis*, vol. 4, no. 3, September 1988, p. 211.

8. Command by personal direction or detailed directives: See Thomas J. Czerwinski, "Command and Control at the Crossroads," *Marine Corps Gazette*, October 1995.

9. Foster, p. 211.

10. Authoritarian (Theory X) versus persuasive (Theory Y) leadership: Douglas McGregor, *The Human Side of Enterprise* (New York: McGraw-Hill, 1960), chapters 3 and 4. Situational Leadership Grid (telling, selling, participating, delegating) and follower maturity: Paul Hersey and Kenneth H. Blanchard, *Management of Organizational Behavior*, 2d ed. (Englewood Cliffs, NJ: Prentice-Hall, 1972), p. 134.

11. Integrated teams (work groups): R. Likert, "The Principle of Supportive Relationships," in Derek Salmon Pugh, comp., *Organization Theory: Selected Readings* (Harmondsworth, England: Penguin Books, 1971), pp. 279-304. Figure 5 adapted from Likert, p. 289.

12. Effective organizations characterized by intense communications: Thomas J. Peters and Robert H. Waterman, Jr., *In Search of Excellence* (New York: Harper & Row, 1982), p. 122.

13. On the relative importance of verbal and nonverbal communication: Psychologist Dr. Albert Mehrabian has estimated that in face-to-face conversation the actual meaning of words accounts for a mere 7 percent of communication, nonverbal voice (such as tone, volume, or inflection) accounts for 38 percent, and visible

signals (facial expression, body language, gestures, etc.) account for the re-
maining 55 percent of the communication that takes place. Albert Mehrabian, *Nonverbal Communication* (Chicago: Aldine-Atherton, 1972), p. 182.

14. Supply-push/demand-pull and "demand-cascade": James P. Kahan, D. Robert Worley, and Cathleen Stasz, *Understanding Commanders' Information Needs* (Santa Monica, CA: Rand Corporation, 1989), pp. 37–55.

15. The effects of uncertainty and time on decisionmaking: John F. Schmitt, "Observations on Decisionmaking in Battle," *Marine Corps Gazette*, March 1988, pp. 18–19.

16. Intuitive (naturalistic) versus analytical decisionmaking: Gary A. Klein, "Strategies of Decision Making," *Military Review*, May 1989, and "Naturalistic Models of C^3 Decision Making," in Stuart E. Johnson, Alexander H. and Ilze S. Levis (eds.), *Science of Command and Control* (Washington: AFCEA International Press, 1988).

17. "Satisfice" versus "optimize": Herbert A. Simon, "Rational choice and the structure of the environment," *Psychological Review*, vol. 63, 1956, pp. 129–138.

18. Intuitive decisionmaking more appropriate for the vast majority of tactical/operational decisions: A 1989 study by Gary A. Klein (based on 1985 observations) estimated that decision makers in a variety of disciplines use intuitive methods 87 percent of the time and analytical methods 13 percent of the time. Evidence now suggests that this study was actually biased *in favor of analysis.*

145

More recent studies estimate the breakdown at more nearly 95 percent intuitive to 5 percent analytical. G. A. Klein, "Recognition-Primed Decisions" in William B. Rouse (ed.), *Advances in Man-Machine System Research* (Greenwich, CT: Jai Press, 1989); G. L. Kaempf, S. Wolf, M. L. Thordsen, and G. Klein, *Decision Making in the Aegis Combat Information Center* (Fairborn, OH: Klein Associates, 1992); R. Pascual and S. Henderson, "Evidence of Naturalistic Decision Making in Command and Control" in C. Zsambok and G. Klein (eds.), *Naturalistic Decision Making*, forthcoming publication (Hillsdale, NJ: Lawrence Erlbaum Associates); Kathleen Louise Mosier, *Decision Making in the Air Transport Flight Deck: Process and Product*, unpublished dissertation (Berkeley, CA: University of California, 1990).

Creating Effective Command and Control

1.   All commanders in their own spheres . . . : Spenser Wilkinson, *The Brain of an Army: A Popular Account of the German General Staff* (Westminster: A. Constable, 1895), p. 106.

2.   Initiative as a source of energy in crisis: Fayol, *General and Industrial Management*, p. 39.

3.   Implicit understanding and communication: Boyd, "An Organic Design for Command and Control," p. 18.

4.   "A good plan violently executed . . .": George S. Patton, *War As We Knew It* (New York: Bantam Books, 1980), p. 335.

5.   Hybrid   information   management: Kahan,   et   al.,

Understanding Commanders' Information Needs, pp. 66-67.

6.   The purpose of procedures "not to restrict human judgment . . .": Richard E. Simpkin, *Race to the Swift: Thoughts on Twenty-First Century Warfare* (London: Brassey's Defence Publishers, 1985), p. 239.

7.   "A bulky staff implies . . .": William T. Sherman, *Memoirs of General William T. Sherman* (New York: Da Capo Press, 1984), p. 402.

www.ingramcontent.com/pod-product-compliance
Lightning Source LLC
Chambersburg PA
CBHW031941190326
41519CB00007B/605